Methods in Enzymology

Volume 384
NUMERICAL COMPUTER METHODS
Part E

METHODS IN ENZYMOLOGY

EDITORS-IN-CHIEF

John N. Abelson   Melvin I. Simon

DIVISION OF BIOLOGY
CALIFORNIA INSTITUTE OF TECHNOLOGY
PASADENA, CALIFORNIA

FOUNDING EDITORS

Sidney P. Colowick and Nathan O. Kaplan

Methods in Enzymology

Volume 384

Numerical Computer Methods

Part E

EDITED BY

Michael L. Johnson

UNIVERSITY OF VIRGINIA HEALTH SYSTEM
CHARLOTTESVILLE, VIRGINIA

Ludwig Brand

JOHNS HOPKINS UNIVERSITY
BALTIMORE, MARYLAND

ELSEVIER
ACADEMIC
PRESS

AMSTERDAM • BOSTON • HEIDELBERG • LONDON
NEW YORK • OXFORD • PARIS • SAN DIEGO
SAN FRANCISCO • SINGAPORE • SYDNEY • TOKYO
Academic Press is an imprint of Elsevier

Elsevier Academic Press
525 B Street, Suite 1900, San Diego, California 92101-4495, USA
84 Theobald's Road, London WC1X 8RR, UK

This book is printed on acid-free paper.

For all information on all Academic Press publications
visit our Web site at www.academicpress.com

ISBN: 0-12-182789-5

PRINTED IN THE UNITED STATES OF AMERICA
04 05 06 07 08 9 8 7 6 5 4 3 2 1

Dedication

To David A.Yphantis, a valued scientist, a wonderful mentor, and a friend.

Table of Contents

Contributors to Volume 384

Article numbers are in parentheses and following the names of contributors.
Affiliations listed are current.

JAMES L. COLE (13), *Department of Molecular and Cell Biology and National Analytical Ultracentrifugation Facility, University of Connecticut, Storrs, Connecticut 06269*

DANIEL J. COX (7), *University of Virginia Health System, Center for Behavioral Medicine Research, Charlottesville, Virginia 22908*

JULIE DAM (12), *Center for Advanced Research in Biotechnology, University of Maryland Biotechnology Institute, Rockville, Maryland 20850*

R. J. DESA (1, 3), *On-Line Instrument Systems, Inc., Bogart, Georgia 30622-1724*

DAVID F. DINGES (10), *Unit for Experimental Psychiatry, University of Pennsylvania School of Medicine, Philadelphia, Pennsylvania 19104-6021*

WILLIAM S. EVANS (4, 6), *Department of Internal Medicine and Department of Obstetrics and Gynecology, The General Clinical Research Center and the Center for Biomathematical Technology, University of Virginia Health System, Charlottesville, Virginia 22908-0735*

LEON S. FARHY (5, 6), *University of Virginia, Charlottesville, Virginia 22908*

CHRISTOPHER R. FOX (6), *Division of Endocrinology and Metabolism, Department of Internal Medicine, University of Virginia Health System, Charlottesville, Virginia 22908*

R. BLAKE HILL (15), *Department of Biology, Johns Hopkins University, Baltimore, Maryland 21218*

MICHAEL L. JOHNSON (4, 6, 9), *Departments of Pharmacology and Internal Medicine, and the Center for Biomathematical Technology, University of Virginia Health System, Charlottesville, Virginia 22908*

BORIS P. KOVATCHEV (7), *University of Virginia Health System, Center for Behavioral Medicine Research, Charlottesville, Virginia 22908*

DOUGLAS E. LAKE (11), *Department of Internal Medicine (Cardiovascular Division), University of Virginia, Charlottesville, Virginia 22903*

MARC S. LEWIS (14), *Molecular Interactions Resource, Division of Bioengineering and Physical Science, Office of Research Services, Office of the Director, National Institutes of Health, Bethesda, Maryland 20892*

GREG MAISLIN (10), *Biomedical Statistical Consulting, Wynnewood, Pennsylvania 19096*

I. B. C. MATHESON (1, 2, 3), *On-Line Instrument Systems, Inc., Bogart, Georgia 30622-1724*

J. RANDALL MOORMAN (11), *Department of Molecular Physiology and Biological Physics, and the Cardiovascular Research Center, University of Virginia, Charlottesville, Virginia 22903*

MICHAEL MULLINS (8), *Mercer University School of Medicine, Memorial Health University Medical Center, Savannah, Georgia 31406*

ERIK OLOFSEN (10), *Department of Anesthesiology, P5Q, Leiden University Medical Center, 2300 RC, Leiden, The Netherlands*

L. J. PARKHURST (3), *Department of Chemistry, University of Nebraska, Lincoln, Nebraska 68588*

MARIO PERUGGIA (8), *Department of Statistics, The Ohio State University, Columbus, Ohio 43210*

MICHAEL M. REILY (14), *Molecular Interactions Resource, Division of Bioengineering and Physical Science, Office of Research Services, Office of the Director, National Institutes of Health, Bethesda, Maryland 20892*

JOSHUA S. RICHMAN (11), *Medical Scientist Training Program, University of Alabama at Birmingham, Birmingham, Alabama 35211*

PETER SCHUCK (12), *Protein Biophysics Resource, Division of Bioengineering & Physical Science, ORS, OR, National Institutes of Health, Bethesda, Maryland 20892*

JUNFENG SUN (8), *Department of Statistics, The Ohio State University, Columbus, Ohio 43210*

PAUL SURATT (8, 9), *Division of Pulmonary and Critical Care Medicine, University of Virginia School of Medicine, Charlottesville, Virginia 22908*

GURUVASUTHEVAN R. THUDUPPATHY (15), *Department of Biology, Johns Hopkins University, Baltimore, Maryland 21218*

HANS P. A. VAN DONGEN (10), *Unit for Experimental Psychiatry, University of Pennsylvania School of Medicine, Philadelphia, Pennsylvania 19104-6021*

JOHANNES D. VELDHUIS (4), *Department of Internal Medicine and Department of Obstetrics and Gynecology, The General Clinical Research Center and the Center for Biomathematical Technology, University of Virginia Health System, Charlottesville, Virginia 22908-0735*

AMELIA VIROSTKO (4), *Department of Pharmacology and the Center for Biomathematical Technology, University of Virginia Health System, Charlottesville, Virginia 22908*

Preface

The speed of laboratory computers doubles every year or two. As a consequence, complex and time-consuming data analysis methods that were prohibitively slow a few years ago can now be routinely employed. Examples of such methods within this volume include wavelets, transfer functions, inverse convolutions, robust fitting, moment analysis, maximum-entropy, and singular value decomposition. There are also many new and exciting approaches for modeling and prediction of biologically relevant molecules such as proteins, lipid bilayers, and ion channels.

There is also an interesting trend in the educational background of new biomedical researchers over the last few years. For example, three of the authors in this volume are Ph.D. mathematicians who have faculty appointments in the School of Medicine at the University of Virginia.

The combination of faster computers and more quantitatively oriented biomedical researchers has yielded new and more precise methods for the analysis of biomedical data. These better analyses have enhanced the conclusions that can be drawn from biomedical data and they have changed the way that the experiments are designed and performed. This is our fifth "Numerical Computer Methods" volume for Methods in Enzymology. The aim of volumes 210, 240, 321, 383, and the present volume is to inform biomedical researchers about some of these recent applications of modern data analysis and simulation methods as applied to biomedical research.

LUDWIG BRAND
MICHAEL L. JOHNSON

METHODS IN ENZYMOLOGY

VOLUME 244. Proteolytic Enzymes: Serine and Cysteine Peptidases
Edited by ALAN J. BARRETT

VOLUME 245. Extracellular Matrix Components
Edited by E. RUOSLAHTI AND E. ENGVALL

VOLUME 246. Biochemical Spectroscopy
Edited by KENNETH SAUER

VOLUME 247. Neoglycoconjugates (Part B: Biomedical Applications)
Edited by Y. C. LEE AND REIKO T. LEE

VOLUME 248. Proteolytic Enzymes: Aspartic and Metallo Peptidases
Edited by ALAN J. BARRETT

VOLUME 249. Enzyme Kinetics and Mechanism (Part D: Developments in Enzyme Dynamics)
Edited by DANIEL L. PURICH

VOLUME 250. Lipid Modifications of Proteins
Edited by PATRICK J. CASEY AND JANICE E. BUSS

VOLUME 251. Biothiols (Part A: Monothiols and Dithiols, Protein Thiols, and Thiyl Radicals)
Edited by LESTER PACKER

VOLUME 252. Biothiols (Part B: Glutathione and Thioredoxin; Thiols in Signal Transduction and Gene Regulation)
Edited by LESTER PACKER

VOLUME 253. Adhesion of Microbial Pathogens
Edited by RON J. DOYLE AND ITZHAK OFEK

VOLUME 254. Oncogene Techniques
Edited by PETER K. VOGT AND INDER M. VERMA

VOLUME 255. Small GTPases and Their Regulators (Part A: Ras Family)
Edited by W. E. BALCH, CHANNING J. DER, AND ALAN HALL

VOLUME 256. Small GTPases and Their Regulators (Part B: Rho Family)
Edited by W. E. BALCH, CHANNING J. DER, AND ALAN HALL

VOLUME 257. Small GTPases and Their Regulators (Part C: Proteins Involved in Transport)
Edited by W. E. BALCH, CHANNING J. DER, AND ALAN HALL

VOLUME 258. Redox-Active Amino Acids in Biology
Edited by JUDITH P. KLINMAN

VOLUME 259. Energetics of Biological Macromolecules
Edited by MICHAEL L. JOHNSON AND GARY K. ACKERS

VOLUME 260. Mitochondrial Biogenesis and Genetics (Part A)
Edited by GIUSEPPE M. ATTARDI AND ANNE CHOMYN

VOLUME 261. Nuclear Magnetic Resonance and Nucleic Acids
Edited by THOMAS L. JAMES

[1] A Practical Approach to Interpretation of Singular Value Decomposition Results

By R. J. DeSa and I. B. C. Matheson

Introduction

Singular value decomposition (SVD) is widely used to analyze data where the experimental response is a function of two quantities; absorbance or fluorescence changes as a function of time and wavelength, for example. Other possibilities include other types of spectral response [circular dichroism (CD), nuclear magnetic resonance (NMR), infrared (IR), etc.], with time replaced by temperature, concentration, voltage, or another quantity. Application of SVD to such three-dimensional (3D) data yields two sets of eigenvectors, one corresponding to time (or some other variable) and the other to wavelength and a set of eigenvalues. It is often possible to deduce the number of significant species, N_s, present by inspection of the eigenvectors and eigenvalues. The value of SVD analysis is that a subsequent global fit to the data need be done only on N_s eigenvectors, typically 2–4, rather than the 200 or more wavelengths normally present in the raw data. The reduction in the scale of the problem leads to a great reduction in the global fit computation time. SVD also has the useful property of separating significant information from the noise in the data set. If only the first N_s eigenvectors are judged as being significant and are used for a fit to a mechanism, then a 3- to 4-fold noise reduction is observed. The noise reduction can also be seen by comparing the original data to data reconstructed from the N_s significant eigenvectors (see later).

The application of SVD to kinetic data where spectral changes are analyzed as a function of time has been comprehensively discussed by Henry and Hofrichter.[1] The topics discussed included the rationale for SVD application to experimental data, the definition of SVD and its history, the statistical aspects of SVD, and the SVD algorithm itself. An up to date discussion of the SVD algorithm may be found in the latest edition of "Matrix Computations."[2] SVD will be regarded as a *black box* for the purposes of the following discussion.

[1] E. R. Henry and J. R. Hofrichter, *Methods Enzymol.* **210,** 129 (1992).
[2] G. H. Golub and C. F. Van Loan, "Matrix Computations." Johns Hopkins University Press, Baltimore, MD, 1996.

The Optimal Graphic Presentation of SVD Results

The purpose of this chapter is to examine several ways of displaying the results of SVD and to suggest a practical, largely graphic, approach to their interpretation. Figure 1 shows the reaction of tyrosine phenol-lyase with aspartate. The data were collected in the laboratory of Professor Phillips at the University of Georgia. There are 201 wavelengths and 305 time increments evenly spaced at 0.016 s. The results of the application of SVD to the data set are shown in Fig. 2. These results are presented as two sets of six charts, arranged to emphasize the relationship between the kinetic (change) information and the related spectral information.

The SVD process is represented as

$$Y = USV$$

where Y is the 3D data array (Fig. 1), U is a matrix of eigenvectors containing the kinetic information, S is a diagonal matrix of eigenvalues, and V is a matrix of eigenvectors containing the spectral information. In Fig. 2 the

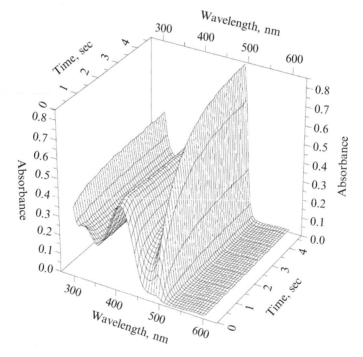

FIG. 1. Three-dimensional plot of the reaction of tyrosine phenol-lyase with aspartate. These data were collected at the laboratory of Professor Phillips at the University of Georgia.

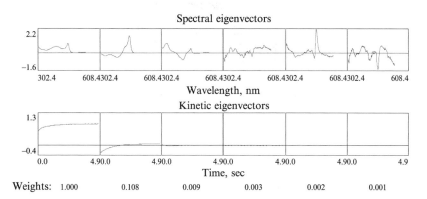

FIG. 2. Application of SVD to the tyrosine phenol-lyase reaction with asparate data of Fig. 1. The spectral data are plotted in the upper six charts as V. The kinetic data are plotted in the lower set of six charts as US.

upper set of six charts contains V and the lower set contains the product US. Each of the upper and lower sets of six charts has been scaled to a common global minimum/maximum. We have found this display where the six kinetic charts underlie the six spectral charts to be the most helpful in visual analysis. The spectral and kinetic effects for each chart are clearly separated from those in other charts. Additionally, the data in any vertically aligned pair of spectral and kinetic charts are correlated. The eigenvalues, displayed under the lower US charts of Fig. 2, have been normalized to their initial value to make it easier to compare different sets of SVD results and are reported as weights. This is the commonly used weighting scheme. It is convenient for fitting since a global kinetic model is fitted to US yielding a set of concentrations varying as a function of time.[3] The purpose of global kinetic fitting is to disentangle from the N_s significant eigenvectors a set of species each with its own time course and spectrum as required by a selected model. The spectra of the N_s species are then obtained by multiplication of the pseudoinverse of the concentration matrix by either the original data Y, or better, a reduced noise data set A. A is constructed as $A = U^1 S^1 V^1$. The superscript 1 signifies that the smaller dimension of the matrices is now N_s, the assumed number of kinetic species present. In mathematical terms N_s is the rank of the matrix necessary to describe the data.

Close inspection of Fig. 2 shows it may not be the optimal presentation for visualizing the data and deciding upon the number of species. The upper charts contain the spectral eigenvectors V that appear to be of comparable amplitude, potentially leading an observer to conclude that all six

[3] E. R. Malinowski, *Anal. Chem.* **49,** 612 (1977).

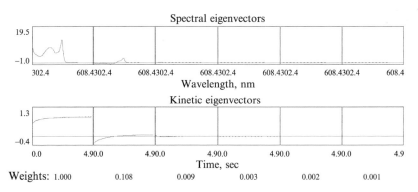

FIG. 3. Application of SVD to the tyrosine phenol-lyase reaction with asparate data of Fig. 1. The spectral data are plotted in the upper six charts as VS. The kinetic data are plotted in the lower set of six charts as US.

V eigenvectors contribute to the data. Also, the US display in the lower charts shows a clear decay shape in the first two charts, a very small change in the third chart, and almost no change in the last three charts, perhaps leading the observer to select $N_s = 2$.

Thus this presentation, we feel, overemphasizes the significance of the spectral eigenvectors and underemphasizes the significance of the kinetic eigenvectors, making assignment of N_s problematic.

We decided to weight the eigenvector data sets differently and assess the utility of the resulting display. Our second attempt was to plot VS and US (Fig. 3), that is, to weight the data equally in the upper and lower sets of six charts. This modified display greatly suppresses the spectral eigenvectors, so that both sets of eigenvectors appear overly suppressed making N_s more difficult to evaluate.

The next weighting scheme we tried was plotting $VS^{1/2}$ and $US^{1/2}$. The results are shown in Fig. 4. The upper and lower charts are still equally weighted and the traces in the charts appear more appropriate. In this case there is content in all six charts in both upper and lower displays. Inspection of the lower kinetic display shows a small degree of noise in charts 1 and 2 and a larger and roughly constant degree of noise in charts 3 to 6. There appears to be significant data in the first three charts. Similarly the upper spectral charts show significant data in the first three charts. There is recognizable content in all 12 charts with none of the eigenvectors seeming out of place or overly suppressed. Observation of the SVD output for many data sets convinces us that the $VS^{1/2}$ and $US^{1/2}$ presentation is the best for enabling a user correctly to assign N_s, and not to be mislead and assign too many species on the basis of spectra in higher V boxes.

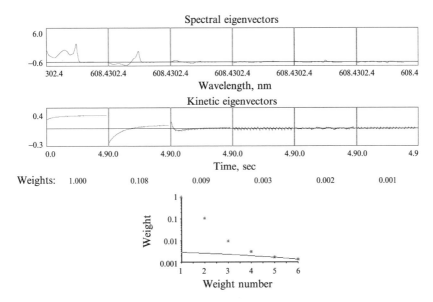

FIG. 4. Application of SVD to the tyrosine phenol-lyase reaction with asparate data of Fig. 1. The spectral data are plotted in the upper six charts as $VS^{1/2}$. The kinetic data are plotted in the lower set of six charts as $US^{1/2}$. The lowermost, centered, plot is a logarithmic plot of weight against weight number.

The $VS^{1/2}$ and $US^{1/2}$ displays have been chosen as standard in the OLIS version of SVD as seen in the general fitting program GlobalWorks. It should be clearly understood that the use of kinetic quantities $US^{1/2}$ and $VS^{1/2}$ is for display purposes only. US is used for fitting and spectra are derived using matrix algebra.

The reader will note that in each of the eigenvector displays (Figs. 2, 3, and 4) a dashed line is drawn through zero in both sets of charts. We find this line is helpful in deciding how many charts contain significant data.

In the case of Fig. 4, the first three charts show clear signs of kinetic change. The next three charts show no sign of kinetic change but periodic spikes at regular intervals. The periodic noise moved into the higher charts by SVD is due to imperfections in the scanning mechanism. SVD is remarkably successful at separating artifacts from the real kinetic effects. The upper spectral charts show significant spectra in the first three charts and very small noisy spectra in the last three. It is clear the last three are not significant. Thus the spectral charts seem less diagnostic than the kinetic charts in this case.

Numerical methods have been used to estimate N_s. We have chosen to use the simple IND method due to Malinowski.[3] When applied to the data of Fig. 4 IND finds $N_s = 3$, in agreement with the visual inspection of the lower kinetic charts. IND is not always in agreement with the visual result; it appears to work best with relatively low noise data.

A graphic aid to N_s selection is to plot the log of the weights against the weight number. The results of this are shown in the lowest panel of Fig. 4. A line joining points 6 and 5 has been extrapolated back to the origin. Three points are clearly above this trend line and suggest $N_s = 3$. Both IND and the log plot of eigenvalues often, but not always, reinforce the selection of N_s based on visual analysis of the charts.

Effect of Noise on Determination of N_S

Synthetic data were generated for the $A \rightarrow B \rightarrow C$ sequential reaction with rate constants of 9 and 3. There were 201 evenly spaced wavelength increments extending from 350 to 550 nm. Spectra were synthesized as unit height skewed Gaussians with maxima at 420, 500, and 550 nm. There were

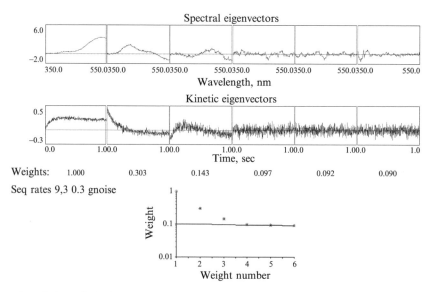

FIG. 5. A synthetic data set for the three species sequential mechanism $A \rightarrow B \rightarrow C$. The spectra used for synthesis were skewed Gaussians with maxima at 420, 500, and 600 nm. The rate constants were 9.0 and 3.0 s^{-1}. Thirty percent of full-scale Gaussian noise was added to the synthesized data. The spectral data are plotted in the upper six charts as $VS^{1/2}$. The kinetic data are plotted in the lower set of six charts as $US^{1/2}$. The lowermost, centered, plot is a logarithmic plot of weight against weight number.

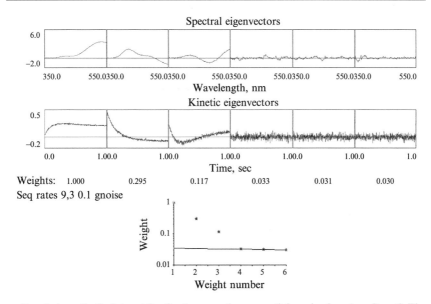

FIG. 6. A synthetic data set for the three species sequential mechanism $A \rightarrow B \rightarrow C$. The spectra used for synthesis were skewed Gaussians with maxima at 420, 500, and 600 nm. The rate constants were 9.0 and 3.0 s^{-1}. Ten percent of full scale Gaussian noise was added to the synthesized data. The spectral data are plotted in the upper six charts as $VS^{1/2}$. The kinetic data are plotted in the lower set of six charts as $US^{1/2}$. The lowermost, centered, plot is a logarithmic plot of weight against weight number.

501 time increments spaced at an even 0.002 s. The results of application of SVD to a synthetic data set with an s/n of 3.3 (full scale) are shown on Fig. 5. The logarithmic plot of the weights is shown in the lowest panel. Both the eigenvectors plot and the weight plot clearly indicate three species are present. However, the IND procedure indicates $N_s = 1$. Another data set was synthesized with an s/n of 10. The results of the application of SVD to this data set are shown in Fig. 6 and the weight plot in the lowest panel. In this case all of the graphic plots indicate $N_s = 3$ but IND still indicates $N_s = 1$. Thus it appears that IND tends to fail on noisy data, while the visual indication is unaffected by noise. It should be noted an $A \rightarrow B \rightarrow C$ sequential fit to both of these synthetic files gives a satisfactory return of the rate constants and spectra used in the synthesis.

Visual Display Indication of Partial SVD Calculation Failure

Figure 7 displays the results of the application of SVD to a data set where the IR spectrum of an unchanging polystyrene infrared filter was observed as a function of time in order to assess machine drift. It is clear that

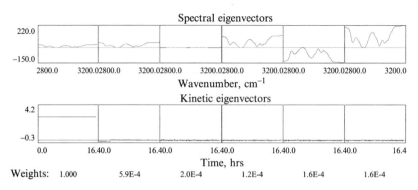

FIG. 7. Results of application of SVD to data where the IR spectrum of a polystyrene filter had been collected as a function of time to assess drift. The spectral data are plotted in the upper six charts as $VS^{1/2}$. The kinetic data are plotted in the lower set of six charts as $US^{1/2}$.

something has gone wrong here. The spectral charts have spectra in the high number charts and the weights are out of their usual monotonic decay sequence. The failure occurs within the Golub–Reinsch algorithm.

Fortunately the occurrence of this failure is rare and is signaled by the eigenvector and eigenvalue displays. When such failure is encountered the results should be regarded with caution. In this particular case the failure is probably associated with the fact that the collected spectrum does not change with time.

Conclusions

The matrix rank or number of species N_s in SVD results may be readily determined by informed visual inspection of an *appropriate* SVD graphic display. The most useful eigenvector display is $VS^{1/2}$ and $US^{1/2}$ with an added line indicating zero on the eigenvector scale. A logarithmic plot of the normalized eigenvectors is also useful. Such eigenvector displays make visual detection of experimental artifacts easy and reliable. SVD will, on rare occasion, fail, usually when there is no change in one of the axes. These failures are signaled by out of sequence eigenvalues and unexpected eigenvectors.

[2] Large Reduction in Singular Value Calculation Time Using Savitzsky–Golay Data Precompression

By I. B. C. MATHESON

Introduction

Devices such as the OLIS rapid scanning monochromator operating in kinetic mode produce three-dimensional (3D) data sets with high hundreds of time increments and low hundreds of wavelength increments. Such data sets consist of an observed quantity, usually absorbance or fluorescence, measured as a function of one or more factors. In the kinetic case these factors are a small number of chemical species each possessing its own decay rate and spectrum. Factor analysis[1–4] has been created to determine how many factors are present in such multivariate data. Since such data sets are often very large, the commonly applied singular value decomposition (SVD) may take an unacceptably long calculation time. The purpose of this work is to introduce a data precompression method that markedly reduces the calculation time for data sets with many wavelengths.

Following the terminology of Malinowski,[1] the data array A is decomposed into an abstract row matrix R and an abstract column matrix C.

$$A = RC \tag{1}$$

Since the number of factors is much less than the total number of points in either axis this is a highly overdetermined system. Many methods have been used for factor analysis. SVD is generally regarded as the method of choice,[2] being more robust and precise than other methods. The algorithm is that of Golub and Reinsch.[2] A recent description may be found in Golub and van Loan.[3] If the data are stored in array A of dimensions m, n SVD is represented by

$$A = USV^T \tag{2}$$

where U is a matrix of eigenvectors of dimensions m, n, and V^T is the transpose of a matrix of eigenvectors of dimensions n, n. U and V are orthonormal

[1] E. R. Malinowski, "Factor Analysis in Chemistry," 2nd Ed. John Wiley & Sons, New York, 1992.

[2] G. H. Golub and C. Reinsch, *Numer. Math.* **14,** 404 (1970).

[3] G. H. Golub and C. H. van Loan, "Matrix Computations," 3rd Ed., Ch. 5 and Ch. 8. Johns Hopkins University Press, Baltimore, MD, 1996.

[4] E. R. Henry and J. Hofrichter, *Methods Enzymol.* **210,** 129 (1992).

matrices such that $\mathbf{UU^T} = \mathbf{I}$ and $\mathbf{VV^T} = \mathbf{I}$, $\mathbf{I}$ being the identity matrix. $\mathbf{S}$ is a diagonal matrix of dimensions $\mathbf{n}$, $\mathbf{n}$ whose values are the square roots of the eigenvalues. It should be noted that some versions of the SVD algorithm return $\mathbf{V}$ and not its transpose $\mathbf{V^T}$. If the diagonal values in $\mathbf{S}$ are inspected it is observed that they start off at a high value and rapidly fall off to a zero or noise value. Similarly visual inspection of plots of $\mathbf{U}$ and $\mathbf{V}$ shows that only a small number have physically significant traces. Comparison with Eq. (1) suggests that $\mathbf{D} = \mathbf{US}$ and $\mathbf{V^T} = \mathbf{R}$. The small number of significant eigenvalues and eigenvector traces represents the number of factors present in $\mathbf{A}$. This is the great advantage of factor analysis, which reduces the data matrix to its lowest dimensionality and yields recognizable factors.[4] The rest of the $\mathbf{U}$ and $\mathbf{V}$ eigenvectors contain noise and may be discarded so that the information in a data set may be stored in a much more compact form. The number of columns in, say, $\mathbf{U}$ is reduced from $\mathbf{n}$ to $\mathbf{o}$, the number of significant eigenvectors. In addition any subsequent data analysis is greatly simplified. The number of significant eigenvectors is typically two to four for real data and rarely as many as six. Thus the data in $\mathbf{A}$ may be safely stored with a dimension 2, $\mathbf{p}$, of columns. The net effect of SVD in highly overdetermined data sets is to extract the information in the large data set and store it in the much smaller arrays $\mathbf{U}$, $\mathbf{S}$, and $\mathbf{V}$. The data may be represented with good accuracy by

$$\mathbf{A} = \mathbf{USV^T}$$

where $\mathbf{U}$, $\mathbf{S}$, and $\mathbf{V^T}$ have a dimension 2 of $\mathbf{p}$. The innovation proposed in this work is to preprocess the data in $\mathbf{A}$ from $\mathbf{n}$ to an intermediate number $\mathbf{p}$ columns using a smoothing routine *before* SVD to determine $\mathbf{o}$, the number of significant factors.

Data Compression Using Savitzky–Golay Smoothing Weights

Savitzky and Golay (SG) published a set of convolution weights for the smoothing (and differentiation) of data.[5] These weights are calculated by direct solution of simultaneous equations and produce results equivalent to least-squares smoothing. The least-squares value for a point is calculated as a weighted combination of the point and the $\mathbf{m}$ points on either side of it, a $(2\,\mathbf{m} + 1)$ point smooth of the data. The SG method is commonly used in the form of tabulated weights that are convoluted appropriately with the data to produce the smoothed result. The advantage of SG is

[5] A. Savitzky and M. J. E. Golay, *Anal. Chem.* **36,** 1627 (1964).

speed relative to direct least squares and the disadvantage for smoothing is that the data are truncated by **m** points at either end. Steiner *et al.*,[6] using a similar method, corrected some errors in the tables published by S. G. Khan[7] and Leach *et al.*[8] have addressed the truncation problem also using a similar method. Gorry[9] addressed the smoothing problem in a different fashion using Gram polynomials and produced an algorithm that calculated the smoothing convolution weights, differentials, and corrected the truncation problem. Thus tables of weights are no longer required and the convolution weights can be calculated as necessary.

The problem here is compression and not smoothing. If a value of **p** is chosen,[8–10] then the nearest $(2\,\mathbf{m} + 1)$ may be calculated as **n2/p**, where **n2** is dimension 2 of **A**. Then going along the data in sequence each $(2\,\mathbf{m} + 1)$ point is replaced by a single point giving a total number of points of **p**. The beauty of the Gorry algorithm for the weights is that it takes a much simpler form if differentials and truncation correction are not required. A program fragment and functions to calculate the SG convolution weights for quadratic smoothing in the programming language FORTRAN are given in the appendix.

It should be noted that the function weight in the appendix takes a very simple nonrecursive form if only the smoothing convolution weights are required. The calculation is relatively fast and takes a negligible time compared to the rest of the SVD process. Also note that in the main fragment only the first **m** + 1 elements of the convolution weights vector **ww** are calculated. The rest can be obtained by reflection since **ww** is symmetric about its center.

The modified SVD procedure, PCSVD, uses the precompression subroutine outlined in the appendix. For comparison the original Golub–Reinsch SVD is named GRSVD. Then the original data **A** are given by **yraw** and this is reduced to **ydata** with a second dimension of 8 to 10 in the data compression. The SVD is executed on the reduced array **ydata** using the standard FORTRAN SVD subroutine. The eigenvalues, s1, are returned from the SVD subroutine. The eigenvalues are normalized to their initial value, i.e., norms = s1(i)/s1(1) and their square roots s2(i) = sqrt[s1(i)] in a loop. The array **umat** is multiplying u in dimension 2 by s1. **Umat** is then inverted with a FORTRAN pseudoinverse subroutine **pinv**. The array **vmat**, equivalent to **V**, is reconstructed by multiplying the

[6] J. Steiner, Y. Termonia, and J. Deltour, *Anal. Chem.* **44**, 1906 (1972).

[7] A. Khan, *Anal. Chem.* **59**, 654 (1987).

[8] R. A. Leach, C. A. Carter, and J. M. Harris, *Anal. Chem.* **56**, 2304 (1984).

[9] P. A. Gorry, *Anal. Chem.* **62**, 570 (1990).

[10] R. J. DeSa and I. B. C. Matheson, *Methods Enzymol.* **384**, 1 (2004).

pseudoinverse of **umat** by the original data **yraw**. The first six sets of **Umat** and **Vmat** are scaled by the eigenvalue square roots **s2** as **kin** = **umat**/**s2** and **spec** = **vmat*****s2**. The results are displayed as **spec**, **kin**, and **norms**. The interpretation of such displays are discussed in more detail in an accompanying paper.[10]

Results

All the data files analyzed are large kinetic experiments with dimension 1 being time with hundreds of increments and dimension 2 being wavelength with typically 201 increments. The modified SVD procedure was written in WATCOM Fortran 77 running under extended DOS and in COMPAQ Fortran 77 and Fortran 90 running under Windows. There are two questions that have to be addressed: (1) Do the modified and original SVD procedures produce the same result? Experimental data for three flashes of a firefly are shown in Fig. 1. The results of application of PCSVD to these data are shown in Fig. 2. A fit to a first order decay is shown in Fig. 3. Note the great improvement in spectrum quality in Fig. 3 as compared to that in Fig. 1. (2) What is the advantage of the modified method, i.e., is it faster?

The Equivalence of Results from Modified PCSVD and Original GRSVD

The question is how does one compare the results of the modified and original SVD procedures? The method used here is to fit the SVD kinetic eigenvectors to an assumed kinetic model using the global kinetic fitting program described in the accompanying publication,[11] and to compare the results for the new and original SVD methods. Real experimental kinetic data have rate constants corresponding to the assumed kinetic model that are not known by definition.

There are possible criteria of goodness of kinetic fit. These are (1) the returned rate constants and their errors, (2) the standard deviation of the fit to the **o** kinetic eigenvectors, and (3) the standard deviation of the fit to the entire data set. The rate constant values and error estimates are of little use in distinguishing between the two SVD methods since they show very little or no difference. The kinetic eigenvector global fit standard deviations are useless because of the scaling value differences described previously. Thus the overall standard deviation, **R**, is the criterion chosen. The kinetic eigenvector returned fit is **K**, and the spectra **W**, so that **R** is defined as

$$\mathbf{R} = \mathbf{KW} - \mathbf{USV^T} \qquad (3)$$

[11] I. B. C. Matheson, L. J. Parkhurst, and R. J. DeSa, *Methods Enzymol.* **384,** 18 (2004).

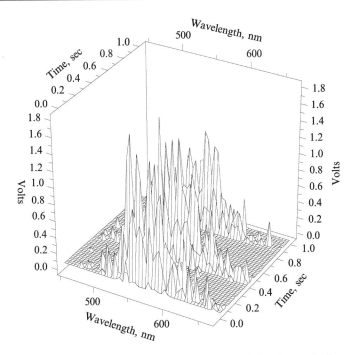

FIG. 1. Bioluminescence spectra: three successive flashes from a firefly.

TABLE I
DATA SETS

File	Fit model	SD, GRSVD	SD, PCSVD
Fire-1	$A \rightarrow \ldots$	2.10e-1	2.10e-1
Fire-3	$A \rightarrow \ldots$	1.51e-1	1.51e-1
Dcipholm	$A \rightarrow, B$	0.361e-2	0.297e-2
Hemo	$A \rightarrow B \rightarrow C$	0.584e-2	0.574e-2
Rssr	$A \rightarrow B \rightarrow C$	1.39e-2	0.985e-2
Dcip	$A \rightarrow, B$	0.662e-2	0.662e-2
43n (synthetic)	$A \rightarrow B \rightarrow C \rightarrow D$	0.124e-2	0.204e-2

Some data sets, real and one synthetic, are compared in Table I and are identified in the appendix.

It is clear that for the real data cases the standard deviation of the overall fit is as good or better for PCSVD in all cases. However, the PCSVD

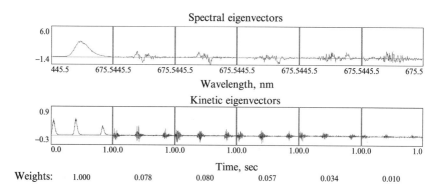

Spectral eigenvectors

Wavelength, nm

Kinetic eigenvectors

Time, sec

Weights: 1.000   0.078   0.080   0.057   0.034   0.010

Fig. 2. Results of application of SVD to the three firefly flashes of Fig. 1. The spectral data are plotted in the upper six charts as $VS^{1/2}$. The kinetic data are plotted in the lower set of six charts as $US^{1/2}$. Both IND and the logarithmic plot of weights indicate only one species is present.

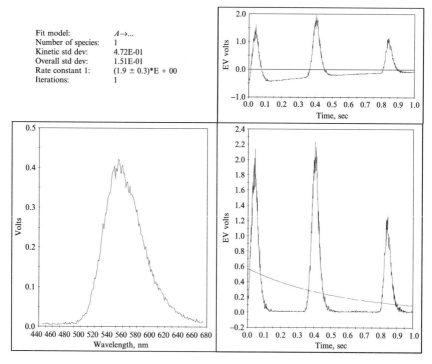

Fit model: $A \to \ldots$
Number of species: 1
Kinetic std dev: 4.72E-01
Overall std dev: 1.51E-01
Rate constant 1: $(1.9 \pm 0.3)*E + 00$
Iterations: 1

Fig. 3. Global fit for a first order decay mechanism, $A \to \ldots$, to the SVD data of Fig. 2. No attempt has been made to fit the shape of the flashes. The simple fit was applied to get a spectrum plot.

method is worse for synthetic data. It is possible that the perfectly Gaussian error distribution of the synthetic data set 4 species 3 rates n is distorted in some way by the compaction process in the PCSVD procedure and causes the poorer fit.

Relative Speed of the New and Original SVD Procedures

It should be noted that the SVD procedure is iterative. Thus it is to be expected that different files will be processed variably since the number of iterations is variable.

Bearing in mind this limitation, timing studies were carried out on the GRSVD procedure as a function of dimension 1 and dimension 2. These showed, to a first approximation, that the time required was proportional to the first power of dimension 1, the number of time increments. Also the times required were proportional to the second power of dimension 2, the number of wavelength increments. Thus also to a first approximation it is predicted that the relative speed increase due to the new SVD procedure will be proportional to a first approximation to the square of the degree of wavelength compression, i.e., nw/9, nw being the number of wavelength increments and 9 being the reduced dim 2 size in both the 201 and 251 wavelength cases.

Table II shows the results for the real data sets analyzed previously. These measurements were made on a 133-MHz Pentium computer running under extended DOS. The time resolution of this system is about 0.05–0.06 s. The PCSVD measurements were for 10 iterations of the algorithm. The time given is that for one iteration. The observed time improvements are a little over two times less than predicted. The differences are easily accounted for by the extra processing the new procedure requires before and

TABLE II
REAL DATA SETS

File	Matrix size	GRSVD time	PCSVD time	Ratio	(nw/9)^2
Dcip	121 × 201	11.31	0.060	189	499
Fire-1	150 × 201	16.64	0.076	219	499
Hemo	401 × 201	45.59	0.185	246	499
Dciphol	401 × 201	47.28	0.187	253	499
Rssr	501 × 201	53.28	0.226	236	499
Fire-3	1000 × 201	118.08	0.456	259	499
43n	501 × 251	95.57	0.284	337	571

after SVD. The pre-SVD processing includes calculating the SG convolution numbers and calculating the reduced dimension 2 = 8 to 10 array. The post-SVD calculations are the inversion of the kinetic information eigenvector array and production of the wavelength information array by multiplication with the original data array.

Conclusions

The new PCSVD procedure is many hundreds of times faster than the original GRSVD procedure on real data arrays of typical sizes. There is no loss of accuracy on real data sets. The last conclusion does not apply to synthetic data with a perfect Gaussian error distribution.

Appendix

```
% Program fragment in MATLAB to carry out data compression and
  SVD.
% Assumes the data set A = yraw(ntimes,nwaves) has been read from a
% file.
%
n = floor(nwaves/m2);
ydata = zeros(ntimes,n);u = zeros(ntimes,n);s = zeros(n,n);v = zeros
  (ntimes,n);s2 = zeros(1,6);
norms = zeros(1,6);umat = zeros(ntimes,6);iumat = zeros(6,ntimes);
  vmat = zeros(6,nwaves);
%
% Data Compression carried here.
for 1 = 1:ntimes
  k = 1;
  for i = 1:n
    ydata(1,i) = 0.0;
    for j = 1:m2
      ydata(1,i) = ydata(1,i) + yraw(1,k)*ww(j);
      k = k + 1;
    end
  end
end
% End data compression and carry out SVD on dimension 2 reduced
  ydata()
%
```

```
[u,s,v] = svd(ydata);
%
for i = 1:6
   norms(i) = s(i,i)/s(1,1);
   s2(i) = sqrt(s(i,i));
end
%
% Create the array umat = u*s2 for display purposes.
for i = 1:ntimes
   for j = 1:6
      umat (i,j) = u(i,j)*s2(j);
   end
end
%
% Invert umat
iumat = pinv(umat);
% Create the array Vmat by reconstruction.
vmat = iumat*yraw;
```

Reaction Identifications

Fire-1: A single flash from a firefly.

Fire-3: Three flashes from a firefly. Note that fire-1 and fire-3 have been fitted to a single exponential decay. This gives an excellent spectrum reconstruction in both cases.

Dcipholmium: Reaction of dicloroindole phenol with ascorbate with a holmium filter in the lightpath.

Hemo: Hemoglobin oxidation by . . . .

Rssr: Reaction of Ellman's reagent: 5:5′-dithiobis(2-nitrobenzoic acid) with excess thioglycerol in a two-step reaction.

43*n*: Synthetic 4 species sequential reaction with rate constants 10, 3, and 1. There were four skewed Gaussian peaks at 280, 370, 425, and 480 nm. The noise added was 5e-3 Gaussian.

[3] Efficient Integration of Kinetic Differential Equation Sets Using Matrix Exponentiation

By I. B. C. MATHESON, L. J. PARKHURST, and R. J. DESA

Introduction

Kinetic data consisting of spectra varying as a function of time are analyzed globally where the determination of the rate constants is linked over all wavelengths. This fitting requires that one or more kinetic differential equations be solved. Such kinetic differential equations arranged in a matrix form may be integrated, i.e., solved, by calculating the exponential of the matrix.

Various methods for the computation of the exponential of a matrix have been known for some time and have been critically reviewed by Moler and Van Loan.[1] Their paper had the intriguing title "Nineteen dubious methods to compute the exponential of a matrix." "Dubious" in this context means varying degrees of mathematical inexactitude.

A system of coupled first-order rate processes can be described by a first-order differential equation in terms of a vector of the "n" chemical species $(\mathbf{X})$ and the $n \times n$ rate-constant matrix $\mathbf{K}$:

$$d\mathbf{C}/dt = -\mathbf{KC} \tag{1}$$

which has a formal solution:

$$\mathbf{C}(t) = \exp(-\mathbf{K}t)\mathbf{C}(0) \tag{2}$$

where $\mathbf{C}(0)$ is the vector of n initial concentrations. It is well known in the physics[2] and linear-systems[3] literature that Eq. (2) can be formally expanded as an exponential series in the matrix $\mathbf{K}t$ (where the scalar t, the time, multiplies each element of the $\mathbf{K}$ matrix) to give a solution for $\mathbf{C}(t)$ [see Eq. (11) and (12) later]. It was proposed[4] and shown that this expansion could be used iteratively to obtain $\mathbf{C}(t)$ and that it could be extended to systems with second-order steps. In this approach, only powers of the $\mathbf{K}$ matrix are required, with no other matrix operations

---

[1] C. B. Moler and C. F. Van Loan, *SIAM Rev.* **20,** 801 (1978).

[2] G. Goertzel and N. Tralli, *in* "Some Mathematical Methods of Physics," p. 21. McGraw-Hill, New York, 1960.

[3] R. Fratilla (ed.), "Theory of Linear Systems," p. 68. Research and Education Association, New York, 1982.

[4] T. M. Zamis, L. J. Parkhurst, and G. A. Gallup, *Comput. Chem.* **13,** 165 (1989).

or determination of eigenvalues of the **K** matrix. This procedure was proposed as an alternative to conventional methods for the numerical integration of coupled differential equations.[5]

On the other hand, a second approach is to apply Laplace transform and various inversion methods to solve (1) and (2) to obtain a closed form solution for $C(t)$ in terms of the eigenvalues of the **K** matrix. The latter approach has been used by one of us in our research.[2,3,6,7] [The "n" eigenvalues can be determined from the roots of an "n" degree characteristic polynomial, with coefficients generated simply from the traces of K^n ($n = 1$ to n) without diagonalization.[8]] The procedure involves carrying out a number of matrix multiplications in setting up the original algebraic expressions and becomes tedious for more than five or so species. A third approach to the problem of solving (1) is well known in the relaxation kinetics literature[9] and is briefly described as follows. The eigenvalues of the $\{rg\}$ relaxation matrix are determined after a symmetrization procedure; the nonzero eigenvalues of the symmetrized matrix $\{r^{1/2} \cdot g \cdot r^{1/2}\}$ are the same as for the **K** matrix. The symmetrized matrix is easily diagonalized by an orthogonal similarity transformation, however, the relaxation matrix is written not in terms of the chemical species, but in terms of the elementary reaction steps. That means that some care must be exercised in handling thermodynamically dependent steps and in transforming from the eigenvectors of the relaxation matrix back to the chemical species to obtain $X(t)$.

A fourth procedure involving diagonalization was described long ago[10] and has been described in some chemical kinetics texts.[11-13] It is briefly presented later to show the close connection to the exponential **K** matrix series approach. It should be pointed out that a number of matrix or equivalent operations are first carried out in all of these latter three methods, compared to the small number of operations required for the exponential matrix series, but once completed, they allow $C(t)$ to be

[5] W. H. Press, B. P. Flannery, S. A. Teukolsky, and W. T. Vetterling, *in* "Numerical Recipes: The Art of Scientific Computing," p. 547. Cambridge University Press, New York, 1989.

[6] Y. Gu, "Hemoglobin: Kinetics of Ligand Binding and Conformational Change." Ph.D. Dissertation, University of Nebraska, 1995.

[7] K. M. Parkhurst, R. M. Richards, M. Brenowitz, and L. J. Parkhurst, *J. Mol. Biol.* **289,** 1327 (1999).

[8] M. Bôcher, *in* "Introduction to Higher Algebra." The Macmillan Co., New York, 1930.

[9] G. W. Castellan, *Ber. Bunsenges. Phys. Chem.* **62,** 898 (1963).

[10] F. A. Matsen and J. L. Franklin, *J. Am. Chem. Soc.* **72,** 3337 (1950).

[11] S. W. Benson, *in* "The Foundations of Chemical Kinetics," p. 39. McGraw-Hill, New York, 1960.

[12] P. C. Jordan, *in* "Chemical Kinetics and Transport," p. 153. Plenum, New York, 1979.

[13] J. W. Moore and R. G. Pearson, *in* "Kinetics and Mechanism," 3rd Ed., p. 296. Wiley Interscience, New York, 1981.

determined for any time without advancing stepwise in time. Although each of these methods could in principle be extended to handle second-order steps in an iterative fashion, it appears that the computational time would be much greater than for the exponential series.

The **K** matrix in general is not symmetric, however, there exists a matrix **T** such that the eigenvectors **Y** can be obtained from, and related to **C** by

$$\mathbf{Y} = \mathbf{TC} \quad \text{and} \quad \mathbf{T}^{-1}\mathbf{Y} = \mathbf{C} \tag{3}$$

Then, since the eigenvectors must obey an eigenvector–eigenvalue equation:

$$d\mathbf{Y}(t)/dt = -\mathbf{\Lambda}\,\mathbf{Y}(t) \tag{4}$$

where $\mathbf{\Lambda}$ is a diagonal matrix of the n eigenvalues; one eigenvalue, corresponding to the equilibrium concentration vector, will be zero. The time evolution of the eigenvectors is then given by

$$\mathbf{Y}(t) = \exp(-\mathbf{\Lambda}t) \quad \mathbf{Y}(0) \equiv \mathbf{E}_t\,\mathbf{Y}(0) \tag{5}$$

[see Eq. (12) later] where $\mathbf{E}_t$ is a diagonal matrix with the **ii**th element $= \exp(-\lambda_i t)$. Making use of Eq. (3),

$$d\mathbf{Y}/dt = \mathbf{T}\,d\mathbf{C}/dt = -\mathbf{TKC} = -\mathbf{\Lambda TC} = (\mathbf{TKT}^{-1}) \cdot \mathbf{Y} \tag{6}$$

and thus,

$$(\mathbf{TKT}^{-1}) = \mathbf{\Lambda} \tag{7}$$

Most matrix diagonalization programs require a symmetric matrix, however, the previously diagonalization can be achieved by obtaining the eigenvalues as described previously[7] with **T** and then $\mathbf{T}^{-1}$ obtained by standard Gaussian elimination. Obtaining the desired solution $\mathbf{C}(t)$ merely involves applying Eq. (3) to Eq. (5) to obtain

$$\mathbf{TX} = \mathbf{E}_t\mathbf{TC}(0) \tag{8}$$

which leads to the desired vector of concentrations as a function of time:

$$\mathbf{C}(t) = \mathbf{T}^{-1}\mathbf{E}_t\mathbf{TC}(0) \tag{9}$$

From Eq. (2), however, we must have

$$\exp(-\mathbf{K}t) = \mathbf{T}^{-1}\mathbf{E}_t\mathbf{T} \tag{10}$$

In the following summation, let the summation be from $n = 0$ to infinity, thus

$$\exp(-\mathbf{K}t) = \sum [(-1)^n/n!](\mathbf{K}t)^n \tag{11}$$

and truncations of this series are thus seen to be approximations to $\mathbf{T}^{-1}\mathbf{E}_t\mathbf{T}$, or, to state the matter somewhat differently, we could show an equivalence between two matrix series, one in $\mathbf{K}t$ and one in $\mathbf{\Lambda}t$ from

$$\mathbf{C}(t) = [\exp(-\mathbf{K}t)]\mathbf{C}(0) = \mathbf{T}^{-1}[\exp(-\mathbf{\Lambda}t)]\,\mathbf{T}\mathbf{C}(0) \tag{12}$$

The matrix series for $\exp(-\mathbf{\Lambda}t)$, however, because of the diagonal structure of $\mathbf{\Lambda}$, is readily shown to be the simple diagonal matrix $\mathbf{E}_t$, since Eqs. (12) and (9) are equal. That is, we do not "need" to expand $\exp(-\mathbf{\Lambda}t)$ in a series, because the result of previous matrix procedures for diagonalizing the $\mathbf{K}$ matrix results in a simple compact matrix that is identical to the infinite series. Stated differently, the effort is placed on initial matrix diagonalizations and multiplications, which then obviate the need for a series expansion. The case for the $\mathbf{K}$ matrix series rests therefore on whether it can be made rapidly convergent with far fewer matrix manipulations than for the other procedures, whether convergence can be obtained even for systems of stiff differential equations, and whether it can be easily extended to include second-order steps. We show later that in fact a very rapid and efficient numerical integrator results from simple modifications of Eq. (11) to yield $\mathbf{C}(t)$ and that the previous criteria are readily met. All further discussion will be concerned with matrix exponentiation, the most direct way of solving Eq. (2).

Matrix exponentiation has been applied to various problems represented by Eqs. (1) and (2). Only a few have been in the chemical kinetic and related fields.

Matrix exponentiation was first applied to model the generation and depletion of isotopes in a nuclear reactor. This is of some importance to chemical kinetics since the secular decay of isotopes is mathematically identical to irreversible sequential first-order chemical reactions. The method is relatively old, the first publication being in 1966.[14,15] There seems to have been a great deal of work on the subject, which may well have driven the interest in matrix exponentiation. A program ORIGEN2, available from Oak Ridge National Laboratory,[16] simulates nuclear fuel cycles and uses matrix exponentiation among other mathematical techniques. It is a relatively large program and can simulate close to 200 isotopes. More recent versions of ORIGEN2 are available.

[14] T. A. Porsching, *Nucl. Sci. Eng.* **25**, 183 (1966).
[15] J. A. Da Nobrega, *Nucl. Sci. Eng.* **46**, 366 (1971).
[16] A. G. Croff, *Nucl. Technol.* **62**, 335 (1983).

The application of matrix exponentiation to chemical kinetics has been much more recent. Zamis et al.[4] showed that kinetic differential equation sets represented by Eq. (1) could be solved using Eq. (2). They showed that the matrix exponentiation method could be applied to reversible and, more importantly, second-order cases, i.e., it was generally applicable to all types of chemical kinetics. Zamis et al.[4] used a series approximation method, which, though not the best method in Moler and Van Loan,[1] used a step size in time that was scaled to the inverse of the trace of $\mathbf{K}$ to achieve convergence. The present work builds upon the work of Zamis et al.[4] by applying the best available matrix exponentiation algorithm. Halvorson[17] has discussed the mathematical basis of matrix exponentiation in this series but did not apply it to examples of real data.

Surprisingly there has been little utilization of matrix exponentiation since Zamis et al.[4] The one field where matrix exponentiation has been employed is the analysis of four-dimensional kinetics, where the observed response is measured as a function of time, wavelength, and temperature.[18] Another case comes from the electrophysiologists, who use conductance measurements to assess the flow of ions through single channels in biomembranes. Some of the kinetics of the conductance follow polyexponential processes and have been modeled using matrix exponentiation.[19,20]

Details of Matrix Exponentiation

The algorithm used is the Padé approximation with scaling and squaring as recommended by Moler and Van Loan.[1] The algorithm used is that of Golub and Van Loan[21] in "Matrix Computations." A double precision subroutine, **expm**, has been written in Fortran 77 and in Fortran 90 by translation of Golub and Van Loan's listing in MATLAB pseudocode.

The first question is what are the contents of the matrix $\mathbf{K}$ to be exponentiated?

Let us consider the case of six independent absorbances or fluorescences, each decaying by a first-order process, i.e., six separate exponential processes. The decay of the six species U–Z may be represented by the following set of differential equations:

[17] H. R. Halvorson, *Methods Enzymol.* **210,** 54 (1992).

[18] A. K. Dioumaev, *Biophys. Chem.* **67,** 1 (1997).

[19] D. Colquhoun and A. G. Hawkes, *in* "Single Channel Recording" (B. Sakmann and E. Neher, eds.), p. 397. Plenum Press, New York and London, 1995.

[20] D. Colquhoun and A. G. Hawkes, *in* "Single Channel Recording" (B. Sakmann and E. Neher, eds.), p. 589. Plenum Press, New York and London, 1995.

[21] G. F. Golub and C. F. Van Loan, *in* "Matrix Computations," p. 572. Johns Hopkins University Press, Baltimore and London, 1996.

$$dU/dt = -k_1 U \qquad \text{(a)}$$

$$dV/dt = -k_2 V \qquad \text{(b)}$$

$$dW/dt = -k_3 W \qquad \text{(c)}$$

$$dX/dt = -k_4 X \qquad \text{(d)}$$

$$dY/dt = -k_5 Y \qquad \text{(e)}$$

$$dZ/dt = -k_6 Z \qquad \text{(f)}$$

The **K** matrix is constructed by rearranging the rate constants in Eqs. (a)–(f) into a square matrix with the species as columns and their derivatives as rows.

$$
\begin{array}{cccccc}
& U & V & W & X & Y & Z \\
dU/dt & -k_1 & 0 & 0 & 0 & 0 & 0 \\
dV/dt & 0 & -k_2 & 0 & 0 & 0 & 0 \\
dW/dt & 0 & 0 & -k_3 & 0 & 0 & 0 \\
dX/dt & 0 & 0 & 0 & -k_4 & 0 & 0 \\
dY/dt & 0 & 0 & 0 & 0 & -k_5 & 0 \\
dZ/dt & 0 & 0 & 0 & 0 & 0 & -k_6
\end{array}
$$

This **K** matrix is notable in that the rate constants corresponding to the depletion of a species lie on the primary diagonal. All other elements are zero. This particular **K** matrix has the useful property that the output of the matrix exponentiation subroutine **expm**, the matrix **EX** is identical to $\mathbf{E}_t$, and thus consists only of the exponentials of the diagonal elements also lying on the primary diagonal. [In Eq. (7), for this case, $\mathbf{T} = \mathbf{T}^{-1} =$ the identity matrix.] Thus the precision of **expm** may be assessed by comparing its output to the results of application of the Fortran supplied function **dexp** to the various rate constants. The results are shown in Table I for rate constants (eigenvalues) that span a range of 10^5. (Stiff differential equations have eigenvalues that span a range greater than 10^2.)

It is clear that the outputs of both **expm** and **dexp** are much more precise than the single precision limit of $\sim$8 significant figures. The difference between **expm** and **dexp** is always less than 10^{-14}. Since experimental data usually have a signal-to-noise ratio of 10^4 or less and are more than adequately represented by single precision arithmetic, it follows that **expm** is more than precise enough to model such data.

The rate constants shown in column 1 of Table I have a range of 10^5. The precision with which the results of **expm** are determined for all of

TABLE I
RATE CONSTANTS

Rate constant	expm	dexp	Difference
10.0000	4.5399908×10^{-5}	4.5399908×10^{-5}	5.5×10^{-15}
1.0000	3.6787942×10^{-1}	3.6787942×10^{-1}	1.8×10^{-15}
0.1000	9.0483741×10^{-1}	9.0483741×10^{-1}	1.2×10^{-16}
0.0100	9.9004983×10^{-1}	9.9004983×10^{-1}	3.4×10^{-16}
0.0010	9.9900050×10^{-1}	9.9900050×10^{-1}	6.7×10^{-16}
0.0001	9.9990000×10^{-1}	9.9990000×10^{-1}	8.9×10^{-16}

the rate constants suggests that the matrix exponentiation algorithm has no problems with stiffness. This is in marked contrast to the oft used Runge–Kutta integration method. A subroutine to implement the Runge–Kutta method would have to have variable step size and would require many steps to achieve an even remotely similar degree of precision. Such a general subroutine would be very difficult to write and extremely slow in execution. The simplicity, precision, and execution speed of **expm** make it by far the preferred method for integrating matrices such as **K**.

Applications to Real Data

Kinetic data are normally collected at a constant time interval. Furthermore, in the absence of second-order steps, the normal case is a **K** matrix corresponding to a set of differential equations containing only first-order rate constants. When both such conditions are met we have a default case where the matrix exponential need be calculated only once no matter how many time increments are present. The initial values of $U–Z$ are used to initialize the **C** vector. The **K** matrix is populated with a set of rate constants given by $k_i = k_i \Delta t$, where Δt is the constant time increment. The subroutine **expm** is used to calculate the matrix **EX** from **K**. Successive values of the vector **C** may be obtained by the series vector matrix multiplication:

$$C_i = C_{i-1} \mathbf{EX} \tag{13}$$

It is convenient to store the complete set of **C** vectors in a matrix **conc**, which has dimensions of **nt** time increments and **ns** species.

Application of Eq. (13) greatly speeds up the calculation of trial models with many time increments. In the interests of minimizing floating point error buildup this vector matrix multiplication is carried out in double precision. Further details of the default and nondefault cases will be given.

Some Notes on OLIS GlobalWorks

Matrix exponentiation is but one component of the OLIS GlobalWorks kinetic data collection and analysis system. Data are collected as a function of time and wavelength. Such data in the matrix $\mathbf{A}$ are subjected to singular value decomposition (SVD) to sets of eigenvectors containing the wavelength and kinetic information, i.e., $\mathbf{A} = \mathbf{USV^T}$.

The number of species, $\mathbf{n}$, is judged from the eigenvector display and the eigenvalues. Details on how this judgment is made[22] and a speed enhancement to the SVD algorithm[23] are given elsewhere in this volume. SVD has two main benefits.

The first benefit is that SVD picks the information content in the data and places it in the first few eigenvectors. Further, the data may be reconstructed by multiplication of the $\mathbf{n}$ eigenvectors and eigenvalues as the noise reduced data matrix $\mathbf{D}$:$\mathbf{D} = \mathbf{U^1S^1V^{T1}}$ in which the "1" superscript denotes a smaller dimension of $\mathbf{n}$ for all matrices.

It is then found that these data are three to four times less noisy than the original data. The data may then be stored as $\mathbf{U}$, $\mathbf{S}$, and $\mathbf{V}$. Thus all the significant information in the data may be stored in a compressed file smaller by up to 30-fold.

The second benefit concerns the reduction in scale of the global fit. An OLIS GlobalWorks data file typically has 201 wavelength points. The output of SVD is two sets of six eigenvector charts and six corresponding eigenvalues. Only six charts are shown since it is unlikely that statistically significant fits can be obtained for more than six species. Let us say the user decides three species are present. Then the global fitting calculation is then reduced by a factor of $201/3 = 67$ by fitting a model to the three kinetic eigenvectors rather than the entire data matrix.

The global kinetic fit uses matrix exponentiation to evaluate the function and a simplex algorithm[24,25] to generate the least-squares best fit to the kinetic eigenvector data. The simplex method is unusual in that it does not require the derivatives of the function with respect to the rate constants. Thus the information to construct an error matrix used to estimate the rate constant errors is not readily available. GlobalWorks generates estimates of the rate constant standard deviations by sampling the simplex contents when a relative change of 0.01 in the sum of squares is encountered from iteration to iteration. Each column of the simplex contains the most recent and older values of a rate constant. A standard deviation is

[22] R. J. DeSa and I. B. C. Matheson, *Methods Enzymol.* **384**, 1 (2004).
[23] I. B. C. Matheson, *Methods Enzymol.* **384**, 9 (2004).
[24] J. A. Nelder and R. Meade, *Comput. J.* **7**, 308 (1965).
[25] M. L. Johnson and L. M. Faunt, *Methods Enzymol.* **210**, 1 (1992).

calculated for each of the n rate constants from the $n + 1$ values in each column. The simplex procedure is then restarted and continued until a convergence limit of 0.00001 relative change in the sum of squares. This procedure yields error estimates within a factor of 2 in either direction from those obtained from Monte Carlo analysis. Following Maeder and Zuberbuhler[26] the kinetic matrix is constructed as **US** and the wavelength matrix is **V**. Only the rate constants are optimized by the simplex algorithm. The various eigenvector amplitudes are obtained by linear algebra,[27] i.e., **Amp** = **pinv(conc) US**, where **conc** is the optimized concentration matrix and **pinv(conc)** is the pseudoinverse of **conc**. GlobalWorks displays the fit to a selected kinetic eigenvector, the rate constants, and their standard deviations. A matrix **W** containing the spectra may be calculated by multiplication of the pseudoinverse of the least-squares estimate of the matrix **conc** by the noise-reduced data set matrix **D**, i.e., **W** = **pinv(conc)D**, and "pinv" is a subroutine to generate the pseudoinverse. We have found the simplex nonlinear fitting engine to be much less prone to finding false minima in global fitting applications than the classic Levenberg–Marquardt method.[28]

Global or multiwavelength data fitting has two advantages over single wavelength fitting. The first is the ability to separate close rate constants. This was first demonstrated by Knutson *et al.*[29] who showed that fluorescent species with very similar decay times could be detected by collecting the results of single-photon counting experiments at several wavelengths and analyzing for two decay times over all of the wavelengths. To mimic this test a data set was synthesized with rate constants of $5 \ s^{-1}$ and $6 \ s^{-1}$ with partially overlapped spectra and added Gaussian noise. The **K** matrix for this case is a subset of the six rate constant case above, i.e.:

$$\begin{bmatrix} -k_1 & 0 \\ 0 & -k_2 \end{bmatrix}$$

The results of this fit to the synthetic data set are shown in Fig. 1. The returned rate constants are identical within error to those used for synthesis. A reduced noise data set **D** was constructed by multiplication of the kinetic and wavelength eigenvectors. A wavelength at which the spectra overlapped, 520 nm, was chosen for single wavelength analysis. This **K** matrix for two independent decays was analyzed globally. The results of the single wavelength fit are shown in Fig. 2. The two rate

[26] M. Maeder and A. D. Zuberbuhler, *Anal. Chem.* **62,** 2220 (1990).

[27] W. H. Lawton and E. A. Sylvestre, *Technometrics* **13,** 461 (1971).

[28] P. R. Bevington, *in* "Data Reduction and Error Analysis for the Physical Sciences," p. 235. McGraw-Hill, New York, 1969.

[29] J. R. Knutson, J. A. Beechem, and L. Brand, *Chem. Phys. Lett.* **102,** 501 (1983).

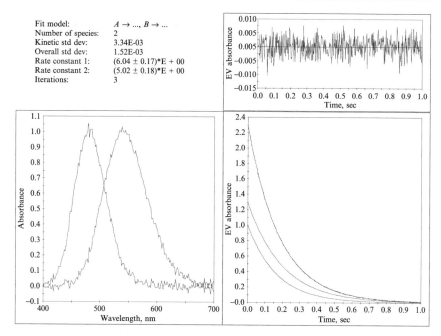

Fig. 1. Global fit to a synthetic data set with two independent first-order fluorescence decays. The first spectrum was synthesized as a skewed Gaussian with a 480 nm wavelength maximum decaying at 6 s^{-1}. The second spectrum was synthesized as a skewed Gaussian with 540 nm maximum decaying at 5 s^{-1}. One percent Gaussian noise was added to the synthesized data. The kinetic standard deviation is that for the kinetic fit to all of the eigenvectors. The overall standard deviation is that for the difference between the returned fit for spectra varying with time and the original 3D data.

constants are not returned correctly in this case, demonstrating the need for global analysis. Note that the residual pattern suggests further complexity.

There is an aphorism in kinetics that kinetics are to mechanism as fact is to fiction. In this case kinetics means rate constants and mechanism is synonymous with the fitting model. This is certainly true in the single wavelength world. However, global measurements and analysis yield the rate constants and time course for each of the **n** species present. Additionally global analysis yields the spectrum of each species. There are cases in which the spectra of some of the species present are known. If the returned spectra from the fit resemble the known spectra then it is quite possible the assumed model is correct. Additionally, the Global-Works package normally is applied to real absorbance data. In that case

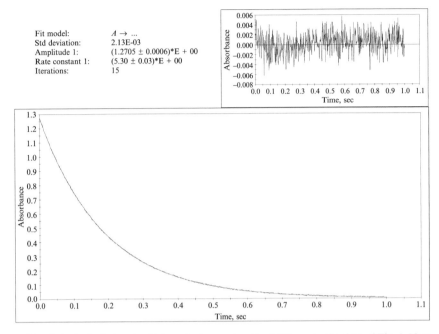

Fit model: $A \rightarrow \ldots$
Std deviation: 2.13E-03
Amplitude 1: $(1.2705 \pm 0.0006)*E + 00$
Rate constant 1: $(5.30 \pm 0.03)*E + 00$
Iterations: 15

FIG. 2. A single first-order fit to a single wavelength of 520 nm to the data of Fig. 1. Note that the returned rate is 5.3 s^{-1}, intermediate between the synthesis rate constants of 6 and 5 s^{-1}. Note also the standard deviation of the fit, 2.13E-03, is significantly worse than the 1.52E-03 standard deviation for the overall fit of Fig. 1.

a fit model that returns spectra with negative value spectra is likely to be incorrect. Further discussion of the importance of spectra will be included in the discussion of the various matrix exponentiation cases.

Applications of Matrix Exponentiation

Sequential Irreversible Reactions

These cases are the most commonly encountered in global fitting with the two to four species cases being the commonest of all. Let us consider the following reaction scheme: $A \rightarrow B \rightarrow C$. This reaction scheme is of some interest since an exact integrated solution is readily obtained. The concentrations of the species A, B, and C at time t, at $t = 0$, $A = A_0$, $B_0 = C_0 = 0$, are given by

$A_t = \exp(-k_1 t)$

$B_t = k_1/(k_2 - k_1) \, [\exp(-k_1 t) - \exp(-k_2 t)]$

$C_t = 1 - k_2/(k_2 - k_1) \, \exp(-k_1 t) + k_1/(k_2 - k_1) \, \exp(-k_2 t)$

The **K** matrix is

$$\begin{bmatrix} -k_1 & 0 & 0 \\ k_1 & -k_2 & 0 \\ 0 & k_2 & 0 \end{bmatrix}$$

The **K** matrix is a compact expression of the set of differentiated equations describing the $A \rightarrow B \rightarrow C$ reaction. The **K** matrix shown is *unique* to this mechanism.

The **EX** matrix resulting from the exponentiation of **K** has the following pattern:

$$\begin{bmatrix} x & 0 & 0 \\ x & x & 0 \\ x & x & 1 \end{bmatrix}$$

The xs in the various elements indicate only nonzero occupation. The x values are generally unequal, and may have any value and sign. The beauty of matrix exponentiation is that the preexponential factors such as $k_1/(k_2 - k_1)$ in the exact solutions preceding are automatically incorporated in the various x elements of **EX**. More complex sequential schemes such as the six species sequential case discussed later have such complex preexponential factors that matrix exponentiation is preferred on grounds of simplicity.

Let us now consider the default case where the time data points are evenly spaced. Then Eq. (3) applies. In this case the vector **C** is replaced by a matrix of allocated size **nt** time increments and three species, i.e., conc(nt,3). Similarly **EX** is allocated as ex(3,3). Application of matrix exponentiation yields an **EX** matrix with nonzero elements on and below the diagonal of the matrix. Then the working Fortran code fragment is

```
do i = 2,nt
   conc(i,1) = ex(1,1) * conc(i − 1,1)
   conc(i,2) = ex(2,1) * conc(i − 1,1) + ex(2,2) * conc(i − 1,2)
   conc(i,3) = ex(3,1) * conc(i − 1,1) + ex(3,2) * conc(i − 2,2) + conc(i − 1,3)
end do.
```

Alternatively, if a Fortran 90 compiler is available the fragment might be

```
do i = 2,nt
  do j = 1,3
    conc(i,j) = sum(ex(1,1 : j) * conc(i − 1,1:j))
  end do
end do.
```

The results obtained for the reaction of xylenol orange (a dye) with ferric ion is shown in Fig. 3A. These data were a gift from Professor Robert Blake (Xavier College, New Orleans). The upper panel shows the eigenvectors and the lower panel the fit. This data set is a good example in that the SVD eigenvectors and the eigenvalues (displayed as weights) both

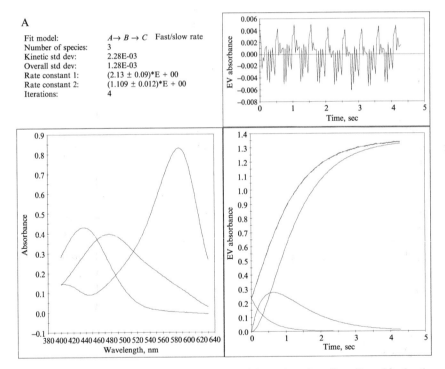

A

Fit model: $A \to B \to C$ Fast/slow rate
Number of species: 3
Kinetic std dev: 2.28E-03
Overall std dev: 1.28E-03
Rate constant 1: $(2.13 \pm 0.09)*E + 00$
Rate constant 2: $(1.109 \pm 0.012)*E + 00$
Iterations: 4

FIG. 3. (A) Global fit of a three-species sequential reaction, $A \to B \to C$, to data for the reaction of xylenol orange with ferric ion. (B) Xylenol orange reaction with ferric ion globally fitted to two independent first-order reactions with a background, $A \to \ldots, B \to \ldots, C$. Note

show very clearly that three species are present. The kinetic fit (lower panel, right side) shows an excellent fit with rather strange looking residuals. This is probably due to the fact that these data were collected with an early version of the OLIS rapid scanning stopped-flow machine that introduced systematic errors. It should be reemphasized that the **K** matrix derived from the differential equations set for the $A \rightarrow B \rightarrow C$ reaction is *unique* to that mechanism.

Consider the simple case of two independent first-order decays in the presence of a background absorbance. The **K** matrix for this is

$$\begin{bmatrix} -k_1 & 0 & 0 \\ 0 & -k_2 & 0 \\ 0 & 0 & 0 \end{bmatrix}$$

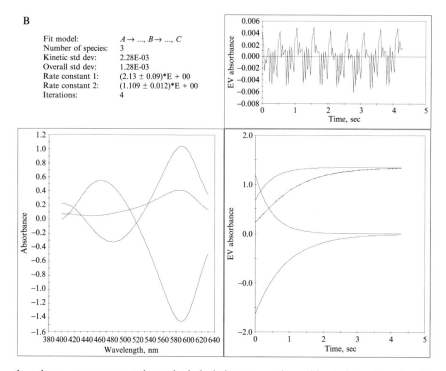

that the rate constants and standard deviation returned are identical to those for the sequential reaction of (A). The negative returned spectra clearly indicate the two independent first-order decays with a background is an unacceptable mechanism.

When the data of Fig. 3A are fitted to this mechanism the results displayed in Fig. 3B are obtained. Clearly the kinetics, i.e., rate constants, are identical. This is true for all eight three-species two-rate constant mechanisms currently offered in GlobalWorks. However, the mechanism fitted in Fig. 3A has all positive spectra suggesting the mechanism is at least plausible. In fact only two of the eight three-species two-rate constant cases give plausible spectra. The other successful fit, not shown, is for $A \rightarrow B \rightarrow C$ with k_1 and k_2 interchanged. Thus it appears matrix exponentiation applied to a *unique* **K** matrix not only obtains the rate constants but also demonstrates highly desirable mechanism selectivity. This is in marked contrast to single wavelength measurements where the rate constants are obtained but the mechanism can only be guessed at.

Matrix exponentiation works just as well for more complex sequential reactions. A case in point is the six-species irreversible sequential reaction for which the **K** matrix is

$$\begin{bmatrix} -k_1 & 0 & 0 & 0 & 0 & 0 \\ k_1 & -k_2 & 0 & 0 & 0 & 0 \\ 0 & k_2 & -k_3 & 0 & 0 & 0 \\ 0 & 0 & k_3 & -k_4 & 0 & 0 \\ 0 & 0 & 0 & k_4 & -k_5 & 0 \\ 0 & 0 & 0 & 0 & k_5 & 0 \end{bmatrix}$$

A data set collected at OLIS for Dr. G. Korshin of the University of Washington is shown in Fig. 4. The data are the chlorination of resorcinol in a probable branching reaction. The mechanism is unknown and from an examination of the eigenvector residuals the application of a six-species sequential reaction model is probably incorrect. Therefore the fact that reasonable looking positive spectra are obtained is fortuitous. The point is to show that five rate constants can be fitted. As before, the left panel shows the spectra and the right the fit. Unlike the preceding $A \rightarrow B \rightarrow C$ reaction, only four eigenvectors in the (not shown) SVD results display seem to contain information. However if the data are fitted with sequential reactions of increasing complexity it is found the six-species fit has the best eigenvector fit and overall standard deviations. Thus the eigenvector display is not an infallible predictor of the number of species. This may be because the weights of the higher eigenvectors approach zero. Further details about the interpretation of SVD displays may be found in Benson.[11]

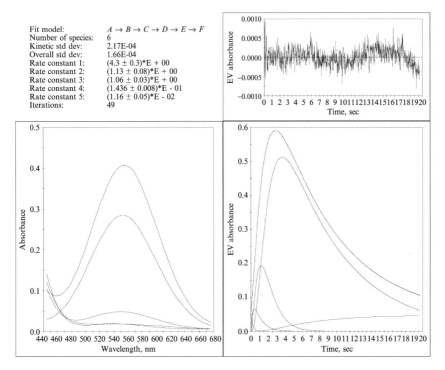

Fit model: $A \to B \to C \to D \to E \to F$
Number of species: 6
Kinetic std dev: 2.17E-04
Overall std dev: 1.66E-04
Rate constant 1: $(4.3 \pm 0.3)*E + 00$
Rate constant 2: $(1.13 \pm 0.08)*E + 00$
Rate constant 3: $(1.06 \pm 0.03)*E + 00$
Rate constant 4: $(1.436 \pm 0.008)*E - 01$
Rate constant 5: $(1.16 \pm 0.05)*E - 02$
Iterations: 49

FIG. 4. A fit to the chlorination reaction of resorcinol. The mechanism is unknown and is possibly branching. The fitting mechanism is a six-species irreversible sequential reaction. Five significant rate constants and six positive, plausible spectra are obtained. The fit is *not* claimed to be the true mechanism. The residual pattern suggests further complexity.

A Second-Order Matrix Exponentiation Case

Zamis et al.[4] demonstrated that matrix exponentiation could be applied to second-order cases. We now show how it may be applied to a simple second order case: $A + A \to B$.

The essential modification is to convert the second-order rate constant elements to pseudo-first-order elements and update the concentration at every time increment. The **K** matrix becomes

$$\begin{bmatrix} -2k_1A & 0 \\ 2k_1A & 0 \end{bmatrix}$$

K(1,2) and **K**(2,2) are set to zero and the initial concentrations set to values of unity (or the true concentrations) and zero. The working code fragment then becomes

```
do i = 2,nt
   K(1,1) = −2k₁ * conc(i − 1,1)
   K(2,1) = −K(1,1)
   call expm(K,EX,2)
   conc(i,1) = ex(1,1) * conc(i − 1,1)
   conc(i,2) = ex(1,1) * conc(i − 1,1) + conc(i − 1,2)
end do.
```

The crucial difference between this working code and that for the first-order cases previously is that the matrix exponentiation subroutine **expm** is called at every iteration rather than only once at the start. Consequently second-order fits take much longer to calculate than first-order ones. The fit to a data set for the dismutation of the superoxide anion is shown in Fig. 5. A second-order kinetic fit has a very characteristic shape, two near asymptotic lines on the Y and X axes connected by a rather sharp "turn" at about 0.25 s. Only the $A + A \rightarrow B$ fit model can fit this data set satisfactorily; the shape cannot be fitted by a combination of exponentials.

Reversible Cases

A synthetic data set and a very complex data set from the earlier work of Zamis et al.[4] will be discussed.

It is well known that analysis of the reversible reaction $A \leftrightarrow B$ (k_1 forward, k_2 backward) will yield an observed rate constant $k_{obs} = k_1 + k_2$. In the absence of knowledge of the equilibrium constant there is no unique solution in this reversible case, an infinity of k_1 and k_2 values adding up to k_{obs} fit the data equally well. This can be resolved if one of k_1 and k_2 or their ratio k_1/k_2 is known.

To test this a two-species reversible data set was synthesized with a forward rate constant of 5 and a backward one of 2. The **K** matrix for the fit model is

$$\begin{bmatrix} -k_1 & k_2 \\ k_1 & -k_2 \end{bmatrix}$$

The results of fitting the data set by a two-species sequential *irreversible* model ($A \rightarrow B$) are shown in Fig. 6. There are two points to note. (1) As expected k_{obs} is returned as $k_1 + k_2$. (2) The spectra for A and B were each

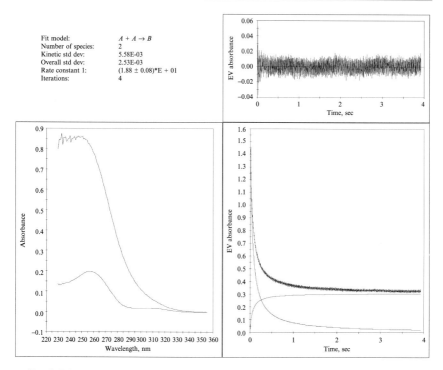

Fit model: $A + A \rightarrow B$
Number of species: 2
Kinetic std dev: 5.58E-03
Overall std dev: 2.53E-03
Rate constant 1: $(1.88 \pm 0.08)*E + 01$
Iterations: 4

FIG. 5. Dismutation of superoxide anion fitted to an homogeneous second-order reaction, $A + A \rightarrow B$. Note that the decay shape is highly characteristic of a second-order reaction and cannot be approximated by combinations of exponentials.

synthesized as skewed Gaussian shapes. The spectrum with the lower wavelength maximum corresponds to species A. It is clear from the appearance of the spectral plot that the second species has apparent structure. Knowing the *synthesized* data spectra have no structure, the fit looks suspect.

This is of limited value, however, since in the real world most spectra have structure. A common criterion for rejection of a model is when the returned spectra have negative components. Such spectra cannot be correct since extinction coefficients cannot be negative. A similar argument holds for rate constants. A further test was to fit the data of Fig. 6 (generated with rate constants $k_1 = 5$ s^{-1}, $k_2 = 2$ s^{-1} and with skewed Gaussians to represent absorption spectra of A and B) to the correct model, $A \leftrightarrow B$, but with rate constants constrained to be incorrect, though close to the correct values. Starting values for pairs of rate constants were 4.5 and 2.5, 4.8

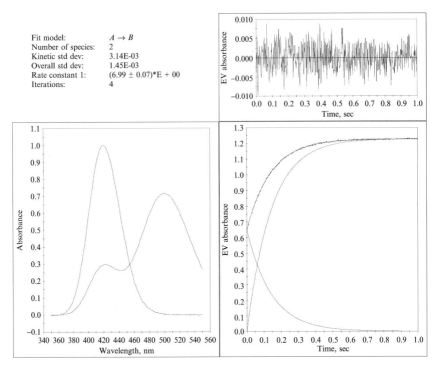

FIG. 6. Global fit to a synthetic data set where a species "A" having a skewed Gaussian spectrum (with a wavelength maximum at 420 nm) reacts reversibly to produce a species "B" having a skewed Gaussian spectrum with a maximum at 500 nm. The forward and backward rate constants were 5.0 s^{-1} and 2.0 s^{-1}. Gaussian noise (1% of the peak value) was added to the synthesized data. The data were fitted to the irreversible two-species sequential mechanism, $A \rightarrow B$. Note that the returned rate constant is the sum of the two rate constants used for synthesis.

and 2.2, 5.0 and 2.0, and 5.5 and 1.5. The fitting program was set to fix k_1 and calculate the best fit for k_2. Note that the sum of the rate constants (7) was that used to generate the data. The spectra and fitting statistics are shown in the four panels of Fig. 7. Panels I and II show a negative spectrum 1 corresponding to species A and are unacceptable. Panel III, produced with the synthesis rate constants, returns the synthesis spectra. Panel IV returns a distorted spectrum 2 akin to that in Fig. 6. The results suggest that in the case of real data one could by trial and error find values of k_1 and k_2 that give a negative spectrum for species A or B. One would then adjust the rate constants until the negative spectrum just disappeared. These values of k_1 and k_2 might be regarded as the "best fit" to the data.

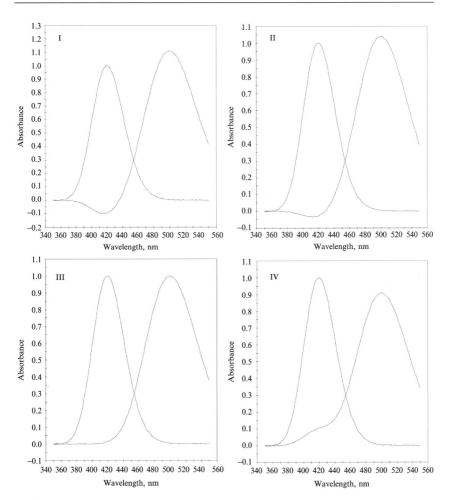

FIG. 7. Spectra produced by four global fits to the data of Fig. 6 by the reversible mechanism $A \leftrightarrow B$. The sum of the forward and backward rate constants was fixed at 7.0 s^{-1}. In all cases the global fits returned exactly the initial values of the rate constants. (Panel I) Rate constant starting values 4.5 s^{-1} and 2.5 s^{-1}. The global fit is clearly unacceptable since one spectrum is negative. (Panel II) Rate constant starting values 4.8 s^{-1} and 2.2 s^{-1}. The results of the global fit are still unacceptable since one spectrum is just negative. (Panel III) Rate constant starting values 5.0 s^{-1} and 2.0 s^{-1}. The synthesized spectra are returned exactly indicating a "true" fit. (Panel IV) Rate constant starting values 5.5 s^{-1} and 1.5 s^{-1}. The global fit is acceptable since positive spectra are returned. These are not, however, the spectra used in the data synthesis.

A Complex Reversible Case

The complex case originally collected and analyzed by Castellan[9] was reanalyzed by Zamis et al.[4] using their version of matrix exponentiation for the following mechanism:

$$
\begin{array}{ccc}
A_1 + A_2 \underset{k_2}{\overset{k_1}{\longleftrightarrow}} A_4 \\
k_6 \updownarrow k_5 \qquad\qquad k_8 \updownarrow k_7 \\
A_3 + A_2 \underset{k_4}{\overset{k_3}{\longleftrightarrow}} A_5
\end{array}
$$

The rather formidable **K** matrix corresponding to this mechanism is

$$
\begin{bmatrix}
X_2 k_1 A_2 + k_5 & X_1 k_1 A_1 & -k_6 & -k_2 & 0 \\
X_2 k_1 A_2 & X_1 k_1 A_1 + X_3 k_3 A_3 & X_4 k_3 A_2 & -k_2 & -k_4 \\
-k_5 & X_3 k_3 A_3 & X_4 k_3 A_2 + k_6 & 0 & -k_4 \\
-X_2 k_1 A_2 & -X_1 k_1 A_1 & 0 & k_2 + k_7 & -k_8 \\
0 & -X_3 k_3 A_3 & -X_4 k_3 A_2 & -k_7 & k_4 + k_8
\end{bmatrix}
$$

where

$$
X_1 = A_1/(A_1 + A_2), \qquad X_2 = A_2/(A_1 + A_2)
$$

$$
X_3 = A_3/(A_2 + A_3) \quad \text{and} \quad X_4 = A_2/(A_2 + A_3)
$$

Zamis et al.[4] showed that the "exact" fits using matrix exponentiation differed by as much as 2% from the usual linearized solutions of relaxation theory. The somewhat awkward problem of dealing with a thermodynamically independent reaction step in the usual relaxation format was completely obviated by use of the **K** matrix. They also showed that the sensitivity coefficients needed for gradient minimization techniques (e.g., Davidson–Fletcher–Powell) are also easily obtained by matrix exponentiation. We present the matrix here to show that complex mechanisms are easily dealt with by matrix exponentiation. Their parameter values are summarized in Table II.

It should be noted that all the comments regarding the uniqueness or otherwise of reversible fits apply only to single experiments at one or many wavelengths. If, for example, global fits to a given reaction are carried out at

TABLE II

PARAMETER VALUES

Rate constant	Value	Concentration	Initial	Eqbm.
k_1	0.50	A_1	9.25	10.00
k_2	1.00	A_2	4.00	5.00
k_3	0.80	A_3	2.75	3.00
k_4	0.40	A_4	25.75	25.00
k_5	0.03	A_5	30.25	30.00
k_6	0.10			
k_7	0.30			
k_8	0.25			

several temperatures it is possible to carry out a grand global fit over the global fits at all temperatures.[18,30] The results of such a fit yield disentangled reversible rate constants and as a bonus their activation parameters.

Conclusions

Matrix exponentiation is an integration method for all kinetic schemes with any degree of complexity. It can be used for the integration of kinetic schemes that are irreversible, reversible, second order, and reactions of such complexity that algebraic integration is impossible. It is ***naturally*** suited to the integration of sets of differential equations that characterize first- and pseudo-first-order chemical reaction networks since it assumes ***exact exponential functions*** as the results. The results of the application of a second-order fit to the data in Fig. 5 suggest the method is also capable of integrating second-order processes. The results of Table I show that matrix exponentiation is more than precise enough for such integrations. This is in marked contrast to numerical integration methods such as Runge–Kutta or Gear where the ***function is approximated***. Last but not least, the demonstrated mechanism selectivity is a powerful tool in ***discrimination*** among various mechanisms.

The simplicity, elegance, precision, and mechanism discrimination of matrix exponentiation make it the method of choice for the integration sets of differential equations sets for diverse mechanisms.

Acknowledgments

This work was supported in part from the following NIH grants to L.J.P.: GM 59346, CA 76049, and RR 15635.

[30] J. F. Nagle, *Biophys. J.* **59,** 476 (1991).

[4]   Deconvolution Analysis as a Hormone Pulse-Detection Algorithm

By MICHAEL L. JOHNSON, AMELIA VIROSTKO, JOHANNES D. VELDHUIS, and WILLIAM S. EVANS

Introduction

Simultaneous estimation of admixed basal and pulsatile secretory activity from serial plasma hormone concentration measurements presents a formidable analytical challenge due to the strong correlations among the key (unobserved) secretion and elimination parameters. However, quantifying nonpulsatile and pulsatile contributions to neuroendocrine signaling is critical to a better understanding of specific regulatory mechanisms. As an initial step toward addressing this issue, we have formulated an automated parametric deconvolution strategy. The recursive algorithm (PULSE4) applies relevant statistical tests to successive estimates of secretory burst mass and basal hormone secretion conditional on iteratively assessed peak positions. Monte Carlo simulations of luteinizing hormone- and insulin-like pulse trains superimposed upon variable basal release disclosed true-positive and false-positive pulse identification rates of 95–100% and approximately 2%, respectively, under optimal conditions. Further analysis showed that the optimal sampling frequency ranges from three to five measurements per hormone half-life and the de facto interpulse intervals exceeds the hormone half-life by 1- to 1.25-fold. A priori knowledge of the imbedded hormone half-life enhanced, but was not required for adequate estimates of unobserved basal secretion. We conclude that iterative least-squares deconvolution analysis of combined basal and pulsatile hormone secretion patterns can be formulated usefully as a peak-identification methodology.

The serum concentrations of many hormones change by orders of magnitude multiple times within each day. Examples of such hormones include luteinizing hormone (LH), growth hormone (GH), prolactin (PRL), thyrotropin (TSH), and adrenocorticotropic hormone (ACTH). Temporal variations in hormone concentrations provide a critical means by which endocrine glands communicate with their remote target organs.[1–4] Thus, it is important to be able to identify and quantify the pulsatile nature

[1] C. Desjardins, *Biol. Reprod.* **24,** 1 (1981).

[2] R. J. Urban, W. S. Evans, A. D. Rogol, D. L. Kaiser, M. L. Johnson, and J. D. Veldhuis, *Endocr. Rev.* **9,** 3 (1988).

[3] R. J. Urban, D. L. Kaiser, E. van Cauter, M. L. Johnson, and J. D. Veldhuis, *Am. J. Physiol.* **254,** E113 (1988).

of such time series of serum concentrations. Many algorithms have been developed for this analysis.[2–5]

In 1987 we noted that the temporal shape of hormone pulses is a convolution integral of secretion into and elimination from the plasma:

$$C(t) = C_0 + \int_0^t S(z)E(t - z)dz \tag{1}$$

where $C(t)$ is the concentration of serum hormone at any positive time, $t \geq 0$; C_0 is the concentration of hormone at the first time point, $t = 0$, $S(z)$ is the amount of hormone secreted at time z per unit time and unit distribution volume, and $E(t - z)$ is the amount of hormone elimination that has occurred in the time interval $t - z$.[6] Consideration of the temporal shape of hormone pulses has provided a significant and different perspective with which to identify and to quantify episodic hormone glandular secretory activity. A relevantly chosen parametric deconvolution process allows one to evaluate the time course of the secretion of glandular products into circulation. The temporal nature of the inferred secretion profiles can subsequently be used to identify hormone pulses and thereby characterize their regulation.

The analysis of data from immunologically based assays, utilized to evaluate hormone concentration time series, poses some unique issues driven by the properties of the experimental uncertainties inherent in the data. For example, hormone assays and attendant sampling procedures are expensive, and thus provide relatively few data points. For various experimental reasons missing data points may limit analysis further by yielding incomplete data sets. In addition, current sampling and assay techniques yield considerable random variability, i.e., experimental noise. Furthermore, the magnitude of such experimental uncertainties is not constant but a complex function of the hormone concentration. Consequently, any method of hormone pulse analysis must explicitly consider (1) sparse time series, (2) missing data, (3) large experimental uncertainties, and (4) variable experimental uncertainties.

Most deconvolution methods were developed for spectroscopic and/or engineering applications where the signal-to-noise ratio is high (i.e., low experimental uncertainties), data series are ample, and observations are equally spaced (i.e., no data points are missing).[7] For physical science

[4] W. S. Evans, M. J. Sollenberger, R. A. Booth, A. D. Rogol, R. J. Urban, E. C. Carlsen, M. L. Johnson, and J. D. Veldhuis, *Endocr. Rev.* **13**, 81 (1992).

[5] J. D. Veldhuis and M. L. Johnson, *Am. J. Physiol.* **250**, E486 (1986).

[6] J. D. Veldhuis, M. L. Carlson, and M. L. Johnson, *Proc. Natl. Acad. Sci. USA* **84**, 7686 (1987).

[7] P. A. Jansson, *in* "Deconvolution with Applications in Spectroscopy." Academic Press, New York, 1984.

applications, the desired result is a continuous spectrum, and, consequently, common deconvolution methodologies are nonparametric. In contrast, for biological applications, we have implemented a parametric approach, wherein one models the secretion, $S(t)$ in Eq. (1), and the elimination, $E(t)$ in Eq. (1), with distinct and physiologically relevant mathematical forms. One then performs a variably weighted least-squares fit of the hormone concentration time series in terms of the parameters for the secretion and elimination equations, $S(t)$ and $E(t)$. Given the nature (i.e., magnitudes and distributions) of the experimental uncertainties contained within the observations, this approach is a maximum likelihood method. An analogous iterative reconvolution technique is utilized in the evaluation of fluorescence lifetimes.[8]

Specifically we assume that the secretion profile can be described as a series of distinct Gaussian-shaped pulses, i.e., secretion events:

$$S(z) = S_0 + \sum_{i=1}^{n} e^{\log H_i - \frac{1}{2}\left(\frac{z - PP_i}{Burst\ SD}\right)^2} \qquad (2)$$

where S_0 is the basal secretion, $\log H_i$ is the logarithm of the height of the ith secretion event, PP_i is the position of the ith secretion event, and $Burst\ SD$ is the standard deviation describing the width of the secretion events. The logarithm is used to constrain the analysis to produce only positive values for the amplitudes of the secretion events, H_i. The full-width at half-height of the secretion events is approximately 2.354 times the $Burst\ SD$. The elimination function is empirically based on both one [Eq. (3)] and two [Eq. (4)] compartment pharmacokinetic models of elimination:

$$E(t - z) = \left(\begin{array}{ll} e^{-\left[\frac{\ln 2\ (t-z)}{HL}\right]}, & \text{if } t - z \geq 0 \\ 0, & \text{if } t - z < 0 \end{array} \right) \qquad (3)$$

where HL is the one compartment elimination half-life and

$$E(t - z) = \left(\begin{array}{ll} (1 - f_2)e^{-\left[\frac{\ln 2\ (t-z)}{HL_1}\right]} + f_2 e^{-\left[\frac{\ln 2\ (t-z)}{HL_2}\right]}, & \text{if } t - z \geq 0 \\ 0, & \text{if } t - z < 0 \end{array} \right) \qquad (4)$$

where HL_1 and HL_2 are the two half-lives of the two compartment model and f_2 is the fractional amplitude corresponding to HL_2. The zeros in the second half of these definitions indicate that elimination does not take place before secretion occurs. In mathematical terms, the elimination is multiplied by a Heaviside function.

[8] J. R. Lakowicz, in "Principles of Fluorescence Spectroscopy." Plenum, New York, 1983.

In our 1987 construction,[6] putative pulse locations are first nominally identified by discrete peak-detection methods, such as CLUSTER analysis.[5] Peaks are manually removed based on failed statistical confidence intervals for the fitted pulse amplitudes. Pulses are manually added based on evident regions of consecutive positive residuals (differences between the observed and predicted reconvolution curves), thus allowing for potential subjectivity.

The present work outlines and tests an expressly automated deconvolution algorithm to analyze hormone concentration time series. We specifically examine the robustness of the operating characteristics of this algorithm in relation to the identification of (true-positive) hormone pulses under expected variance constraints and in relation to a priori known versus unknown hormone half-lives.

Methods

Algorithmic Outline

By way of overview, the automated algorithm, PULSE4, initially generates an expanding series of presumptive peak locations, and then successively removes those that do not meet statistical significance criteria. The primary objective is to evaluate the secretion event positions, PP_is, and amplitudes, $\log H_i$s. To create testable peaks, a natural logarithmic transformation is performed on the data values minus the "Initial Basal Concentration" (later). Data values less than the "Initial Basal Concentration" are discarded at this first stage, and only this stage, of the algorithm.

The logarithmically transformed data can be approximated to some degree as a series of discontinuous straight lines. The locations of the discontinuities correspond to the approximate locations of the secretion events, and the slopes of the predicted lines are related to the apparent single component elimination half-life. Consequently, the initial portion of the algorithm finds the most likely places to put these discontinuities by performing a series of weighted linear least-squares fits, first assuming that there are no secretion events, i.e., that the data are described by an initial concentration at time zero and a decay time.

The algorithm now locates the best location to insert the first presumptive secretion peak by evaluating the statistical consequences of inserting a single peak at every data point location (i.e., time value). For each possible location of a secretion event a weighted linear least-squares fit is performed. The resulting variance of fits are tabulated for each tested secretion-event location. The first presumptive secretion-event location is assigned based on the location yielding the lowest variance of fit associated with a positive secretion amplitude. If there are N data points, the first step thus requires N separate weighted linear least-squares parameter estimations.

The foregoing process is repeated to evaluate the locations of the subsequent presumptive peak locations. This process is repeated for three, four, ... presumptive peaks until the residuals appear random by three statistical tests; (1) the Runs Z score ($Z = 2.0$) for too few runs is not statistically different from zero, (2) the insertion of an additional peak does not significantly ($p < 0.05$) lower the variance of fit, and (3) the lag = 1 autocorrelation of the residuals is not significantly positive ($p < 0.05$).

At this stage, the algorithm determines whether the first data point corresponds to a new secretion event location or simply represents the tail of a previous secretion event. If the timing of the first presumptive secretion event lies within three times the "minimum secretion SD" (later), then the concentration at time zero is assigned a constant value of zero. Otherwise, the concentration at time zero is assumed to be a variable to be estimated by the subsequent weighted nonlinear least-squares (WNLLS) procedures.

To test whether the amplitudes of the individual presumptive secretion events are significant, the algorithm calculates the apparent secretion curve at intervals of one-third the minimum time between the data points using the "minimum secretion event *Burst SD*" (discussed later). Secretion events are tested by scanning the calculated secretion curve for peaks (maxima) separated by nadirs (minima). The actual *Burst SD* of the secretion event is then calculated as the average full-width at half-height of the set of identified secretion events divided by 2.354.

The final phase of the PULSE4 algorithm is the parameter refinement section wherein the statistical significance of the remaining presumptive secretion events is evaluated. For the Monte Carlo simulations presented in this report, the refinement stage proceeds by assuming either (1) that the elimination process is a single elimination function as given by the initial value of the elimination half-life determined from the slope of the line segments fitted to the logarithmic transformed data, or (2) by using the one- and two-component elimination half-lives that were used to simulate the time series. Subsequent weighted nonlinear least-squares fitting during the refinement stage is done by a procedure that is mathematically identical to our original multiparameter DECONV program.[6] In the parameter refinement stage, all of the original non-logarithmic transformed data is utilized for the iterative parameter estimations.

For each stage in the refinement process, the algorithm performs an additional series of variably weighted least-squares parameter estimations.

If an error is detected in any of these steps, then the corresponding event is removed and the refinement process is repeated from the beginning. The specific iterative sequence entails the following:

1. Estimation of secretion-event positions only (i.e., the PP_is) while holding the other parameters constant.
2. Estimation of secretion-event amplitudes only (i.e., S_0 and the log H_is) while holding the other parameters at the values found in step 1.
3. Testing whether including the smallest secretion event provides a statistically significant decrease in the overall variance of fit.
4. Estimation of only the secretion-event amplitudes and the single component elimination half-life given the results of step 2, i.e., S_0, log H_is, and the HL; and,
5. Retesting whether the smallest secretion event provides a statistically significant decrease in the overall variance of fit.

The foregoing sequence of parameter estimations is arbitrary but empirically satisfactory. The following error tests are imposed at one or more of the five refinement steps. If a peak position falls outside the actual range of the data in step 1, then it is removed as a presumptive secretion event and the refinement procedure restarts from the beginning. Second, if peaks are coincident (i.e., during the peak position evaluation steps one or more of the peaks has moved to within 2.5 times the secretion event SD of another peak), they are combined into a single secretion event and the refinement procedure is restarted from the beginning. And, third, to determine if any of the secretion events are actually minor shoulders on other secretion events, PULSE4 calculates the current secretion profile at each of the data points. If a nadir (minima) does not occur between consecutive presumed secretion events then the pair of events is replaced with a single event corresponding to the maximum location and the refinement procedure restarts from the beginning.

Steps 3 and 5 determine if the secretion event with the smallest amplitude peak is statistically significant. This is done by temporarily removing it and repeating the parameter estimation to determine the apparent variance of fit. An F test for an additional term is performed to determine if inclusion of the peak provided a statistically significant ($p < 0.05$) decrease in the variance. Since each secretion event adds two parameters (the amplitude and the position of the peak), the number of degrees of freedom for this F test is 2 and the number of data points minus the total number of parameters.

To accept the final solution, as outlined previously, one of two conditions must be met. If either (1) a runs test for too few runs within the residuals ($p < 0.05$) is not significant or (2) the lag $= 1$ autocorrelation of the residuals is not significantly positive ($p < 0.05$) then the solution is acceptable. Ideally the solution should pass both of these tests and the algorithm performs even better when both conditions are met. If both of these tests fail then the entire algorithm is automatically restarted with the Initial Basal Concentration increased by 10%. Finally, (3) this last operation is performed a maximum of six times to eliminate the possibility of an infinite loop.

Monte Carlo Simulations

The algorithm was tested by a series of Monte Carlo simulations. One hundred synthetic data sets were generated for each of the simulations, based upon reasonable estimates for the various typical parameters for mid-luteal phase LH and insulin secretion, as shown in Table I. In these simulations, the Interpulse Intervals and log H_i were assumed to be Gaussian distributed with the means and SD values given. However, the interval from one secretion event to the next was assumed to be positive and as a consequence negative values in the simulation of the Interpulse Intervals were truncated. Once a series of secretion events and amplitudes was simulated, the concentration as a function of time was determined from the convolution integral [Eq. (1)].

TABLE I
TYPICAL VALUES OF THE PARAMETERS USED IN THE SIMULATIONS

	Mid-luteal LH	Insulin
Number of data points	144	150
Time between data points	10 min	1 min
Number of replicates	2	2
Interpulse interval	91 ± 18 min	12 ± 2 min
log H_i^a	0.4 ± 0.2	6.0 ± 0.8
Burst SD[b]	4.5 min	1.0 min
S_0^c	0.04	6.0
Single component *HL*	55 min	3.1 min
Two component HL_1	18 min	2.8 min
Two component f_2	0.37	0.28
Two component HL_2	90 min	5.0 min
Noise SD (minimum)	0.1	0.2
Noise CV	5.0%	3.6%

[a] log H_i is the base 10 logarithm of the secretory pulse amplitude (units of concentration per minute).
[b] *Burst SD* represents the secretory pulse full-width at half-height in minutes divided by 2.354.
[c] S_0 denotes the basal secretory rate (units of concentration per minute).

For each simulation, a realistic level of Gaussian distributed pseudo-random noise was added to the data sets. The variance model for this noise as a function of the simulated hormone concentration is given by

$$Variance\ of\ noise\ (concentration) = [noise\ SD]^2$$
$$+ \left[\frac{concentration\ [noise\ CV\%]}{100}\right]^2 \quad (5)$$

The *noise SD* term is included so that the simulated experimental uncertainties approach a realistic nonzero minimum value as the simulated concentration approaches zero. Whereas these simulations will not exactly duplicate the actual observed mid-luteal LH and insulin concentration time series, they provide a set of realistic profiles to test algorithmic performance under a variety of conditions. Since the "answers" are known for each of these simulations they can be utilized to test the performance characteristics of the algorithm.

Data sets were simulated based upon typical values for the secretion event interpulse intervals, secretion event amplitudes, basal secretion, elimination half-lives, and with realistic pseudo-random "experimental observational errors" added. It is clearly impossible to present all possible permutations of the interpulse intervals, amplitudes, basal secretion, elimination half-lives, and "observation errors." Thus, the approach taken here is to present analyses across a plausible range of perturbations.

Results

Figure 1 illustrates a typical simulation of a mid-luteal LH concentration (± 1 SEM) time series based upon the values presented in Table I. The corresponding secretion profile, as evaluated by the PULSE4 algorithm with an unknown single elimination half-life, is shown in Fig. 2. Figure 1 also shows the calculated LH concentration profile (reconvolution curve) based upon the inferred secretion profile.

The computational output of the PULSE4 algorithm is not actually a continuous curve as shown in Fig. 2, but rather a table of secretion event locations, PP_is, and the logarithms of the corresponding secretion event heights, log H_is. Figure 2 is simply a calculated secretion curve based upon the final table of pulse-parameters values estimated by the PULSE4 algorithm.

The particular example shown in Figs. 1 and 2 was chosen to demonstrate some difficult to locate secretion events that were located by the PULSE4 algorithm. The four diamonds in Fig. 1 correspond to the time locations of

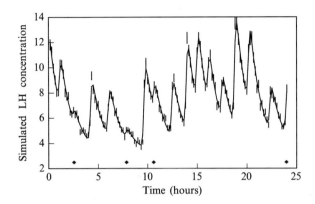

FIG. 1. The vertical lines represent ±1 SEM error bars for a simulated human mid-luteal phase LH concentration time series based upon the typical parameters presented in Table I. The smooth curve in this figure corresponds to the estimated concentration time series based upon the calculated secretion profile shown in Fig. 2, as evaluated by the PULSE4 algorithm assuming an unknown single elimination half-life. The diamonds correspond to the time locations of difficult to locate simulated secretion events illustrated in this analysis.

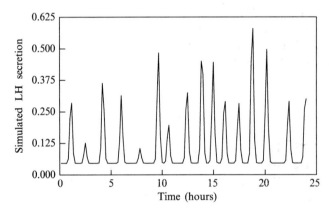

FIG. 2. LH secretion profile inferred from the simulated data shown in Fig. 1 by the PULSE4 algorithm.

four difficult to locate simulated secretion events located by the PULSE4 algorithm in this particular set of data. For example, the fourth diamond marks a simulated peak at 1437.6 min, only 2.4 min before the last simulated data point. The other three marked secretion events are on the trailing edge of a previous secretion event. Sixteen peaks were simulated in this example, and all were located accurately by the PULSE4 algorithm.

TABLE II

CLUSTER AND PULSE4 ALGORITHMIC ANALYSES OF SIMULATED *LH* DATA SETS[a]

CLUSTER algorithm	Simulated single *HL* data	Simulated dual *HL* data
True-positive rate (%)[b]	91.0	92.4
False-negative rate (%)[c]	9.0	7.6
False-positive rate (%)[d]	0.6	0.4
Peak position error (min)	10.1 ± 4.4	8.7 ± 3.5

PULSE4 algorithm	Unknown *HL*	Known *HL*	Unknown *HL*	Known *HL*s
True-positive rate (%)	95.5	99.3	96.7	99.2
False-negative rate (%)	4.5	0.7	3.3	0.8
False-positive rate (%)	0.4	0.8	1.2	1.4
Basal secretion (S_0)	0.054 ± 0.025	0.033 ± 0.003	0.064 ± 0.027	0.035 ± 0.026
HL (min)	48 ± 12		38 ± 10	
Peak position error (min)	0.7 ± 3.3	0.5 ± 1.7	-0.6 ± 2.5	0.4 ± 1.6
Peak mass (observed-true)	0.082 ± 0.105	$0.092 \pm 0.0.081$	0.002 ± 0.094	0.081 ± 0.101

[a] Values are the mean for 100 simulated series $\pm$ SD. See Table I for simulation parameters. *HL* denotes the apparent half-life embodied in the $E(t - z)$ kinetic function [Eqs. (3) and (4)].

[b] The true-positive rate is the percentage of simulated peaks that was located to within two data points by the algorithms.

[c] The false-negative rate is the percentage of simulated peaks that was not located to within two data points by the algorithms.

[d] The false-positive rate is the percentage of peaks located by the algorithms that did not correspond to a simulated peak.

To appraise peak-detection performance, we tabulated true-positive peak identification and false-positive errors as discrimination indices on simulated data series. For example, 100 data sets were simulated using the nominal single elimination LH values in Table I and a second set of 100 data sets was simulated based upon the nominal two component elimination half-lives. The results of the analysis of each of these data set with known and unknown half-lives are summarized in Table II. For example, when the PULSE4 algorithm was applied to the single elimination simulated data under the assumption that the elimination function corresponds to a single unknown half-life, the true-positive rate was 95.5%, the false-negative rate was 4.5%, and the false-positive rate was 0.4%. For comparison, when the same simulated data sets were analyzed with the CLUSTER algorithm,[5] the resulting true-positive rate was 91.0%. The CLUSTER algorithm provides no information about the apparent elimination half-life or basal secretion. The simulated basal secretion rate was 0.04 and the

PULSE4 estimated value was 0.054 ± 0.025 (SD). Uncertainties are represented as SDs that provide a better description of the expected variability for an individual data set than do SEMs. The simulated elimination half-life was 55 min and the estimated value by the PULSE4 algorithm was 48 ± 12 (SD). The mean error in locating a secretion event position was 0.7 ± 3.3 (SD) mins. Corresponding results for the CLUSTER method averaged 10.1 ± 4.4 mins. In judging the error in locating a secretion event remember that the data were simulated at 10-min intervals.

The PULSE4 algorithm can also evaluate the apparent mass of hormone released within each secretion event. The area under a Gaussian-shaped secretion event is

$$Burst\ Area_i = \frac{SD\ 10^{\log H_i}}{\sqrt{2\Pi}} \tag{6}$$

For the present simulation, the average simulated burst mass was 1.96 units. In this setting, the average error (i.e., the difference between the apparent and the simulated burst masses) in burst mass of true-positive peaks was 0.08 ± 0.10 (data were simulated as a single elimination half-life and analyzed by the PULSE4 assuming a single unknown half-life).

One question of particular interest is how often experimental observations should be collected. This question is of immediate practical significance to the experimentalist. For example, being able to utilize a 20-min instead of 10-min sampling frequency can represent substantial research savings. Figure 3 presents the true-positive and false-positive rates as a function of sampling interval for simulated mid-luteal LH data analyzed by the PULSE4 algorithm. It is interesting to note that for long sampling intervals the true-positive rate decreases and for short sampling intervals the false-positive rate increases. For the particular choice of the parameters used in the mid-luteal LH simulation model, the maximum adequate sampling interval appears to be approximately 15 min while the minimum sampling interval appears to be approximately 10 min. However, the important general concept here is that the minimum number of data points per elimination half-life is approximately four, when secretory-pulse detection is the primary outcome of interest.

Figure 3 indicates that at least for this simulation, the true-positive rates decrease as the sampling interval increases and the false-positive rates increase as the sampling interval decreases. However, these changes do not occur as a uniform change. For example, for the 20-min sampling with two known half-lives a total of 223 false-negatives were identified for the 100 simulated data sets. Forty-five of the simulations found zero false-negatives giving a mode of zero. Nineteen of the simulations found

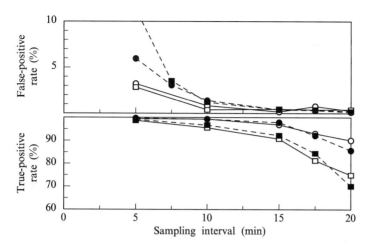

Fig. 3. Apparent true-positive and false-positive rates for the PULSE4 algorithm when applied to mid-luteal phase LH concentration data simulated as described in Table I. The open symbols and solid lines correspond to data simulated for a single-elimination model. The closed symbols and the dashed lines correspond to a two-exponential model. The circles denote PULSE4 analyses based on known elimination half-lives and the squares are for analyses with an unknown single elimination half-life.

only a single false-negative. More than 50% of the false-negatives were from only 11% of the simulations.

The general simulation and testing procedure outlined previously can be utilized to address a number of questions about the sensitivity of the PULSE4 algorithm that cannot be tested directly *in vivo*. For example, Fig. 4 presents an analysis of the sensitivity of the PULSE4 algorithm to a varying interpulse interval of hormone release. For each of the points in Fig. 4, the mid-luteal phase LH parameters shown in Table I were utilized except that the interpulse interval was varied from 40 to 120 min (i.e., from approximately 50% of a half-life to approximately two half-lives). Simulation results indicated that the lower bound of optimal pulse detection is an interpulse interval a little more than one half-life. It is also interesting to note that the PULSE4 algorithm with known half-lives always performs better than PULSE4 with an unknown half-life and both perform better than the CLUSTER method.

The operating characteristics of the PULSE4 algorithm were also explored in relation to other pulsatile hormones. For example, Table III presents the performance data based on 100 simulated insulin data sets reflecting the typical insulin secretory parameters shown in Table I (i.e., the average secretion event mass for this insulin simulation was 2.4).

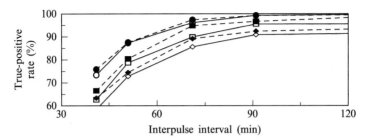

FIG. 4. Apparent true-positive rates for the PULSE4 and CLUSTER algorithms applied to simulated mid-luteal LH phase data of varying mean interpulse intervals. The open symbols and solid lines correspond to data simulated for a single-exponential elimination model. The closed symbols and the dashed lines correspond to data simulated for a two-exponential model. The circles reflect results of PULSE4 analysis with known elimination half-lives, the squares PULSE4 analysis with an unknown single elimination half-life, and the diamonds the CLUSTER algorithm.

TABLE III

COMPARISON OF THE CLUSTER AND PULSE4 ALGORITHM APPLIED TO THE ANALYSIS
OF SIMULATED INSULIN DATA SETS[a]

CLUSTER algorithm	Simulated single HL data	Simulated dual HL data
True-positive rate (%)	92.3	91.5
False-negative rate (%)	7.7	8.5
False-positive rate (%)	0.2	0.7
Peak position error (min)	1.44 ± 0.51	1.46 ± 0.55

PULSE4 algorithm	Unknown HL	Known HL	Unknown HL	Known HL
True-positive rate (%)	99.1	99.5	99.7	99.3
False-negative rate (%)	0.9	0.5	0.3	0.72
False-positive rate (%)	1.0	1.0	0.4	1.4
Basal secretion (S_0)	5.21 ± 0.75	5.85 ± 0.076	5.33 ± 0.80	5.95 ± 0.11
HL (min)	3.47 ± 0.35		3.82 ± 0.42	
Peak position error (min)	−0.097 ± 0.217	−0.002 ± 0.174	−0.11 ± 0.22	−0.01 ± 0.19
Peak mass (observed-true)	−0.124 ± 0.189	−0.072 ± 0.177	−0.155 ± 0.176	−0.076 ± 0.192

[a] Simulation parameter choices are given in Table I.

A comparison of Tables II and III indicates that the PULSE4 algorithm performs better for insulin than for mid-luteal LH. The origin of this improvement is in the distribution of peak sizes. The distribution of the insulin secretion event amplitudes, log H_i, is such that small secretion events do not occur in the insulin simulations. However, for LH a significant number of very small peaks is simulated, which imposes lower true-positive rates.

Conclusions

The present analyses describe and implement an iterative parametric deconvolution technique to identify and quantitate basal and pulsatile hormone release, while concurrently estimating endogenous hormone half-lives. Simulated examples of the application of this methodology indicated that at least for plausible synthetic time series, this parametric deconvolution approach is superior to the CLUSTER method (i.e., one of the commonly accepted procedures) in locating hormone secretion events. PULSE4 also provides crucial information about secretion-event burst mass, elimination half-lives, and basal secretion.

In general, it is very difficult to separate (i.e., deconvolve) the 3-fold contributions to hormone concentration data of pulse amplitudes, basal secretion, and elimination half-lives. This is, in essence, the quintessential ill-posed problem in numerical analysis. However, the present simulations illustrate that a parametric deconvolution approach conditional on iterative reappraisal of peak positions can provide a good estimate of basal secretion without any prior knowledge of the elimination half-lives, and offers an excellent approximation of the amount of basal secretion when the half-lives are known.

The accompanying simulations also offer guidelines about the features of the experimental data that are required for an accurate characterization of the pulsatile nature of hormone release. Specifically, the sampling frequency should be such that there are approximately four or more data points per elimination half-life. In the simulated LH case with a single component elimination half-life of about 55 min the true-positive rates dropped rapidly when data points were simulated less often than every 15 min and the false-positive rates increased when the data were sampled more often than every 10 min (Fig. 3). For insulin with a single component elimination half-life of about 3.1 min, the minimum sampling rates appeared to be about 1 min (data not shown).

Parametric deconvolution-based hormone pulse identification methods also have specific limitations in the maximum detectable hormone pulse frequency. After each secretion event, the concentration of the hormone begins to decrease according to the elimination kinetics. To accurately parameterize the elimination process and basal secretion, a significant amount of elimination must take place before the next secretion event. In general, the minimum time between secretion events needs to equal or exceed 1.25 elimination half-lives (Fig. 4).

Although the present analyses have not explored operating characteristics of pulse detection algorithms for all possible hormones, the principles implied by investigating LH and insulin time-series should be applicable to other neuroendocrine systems. In addition, the foregoing results imply that accurate prior knowledge of hormone-specific kinetics aids significantly in true-positive pulse detection. Likewise, iterative peak analysis conditional on recurrently tested pulse positions avoids subjective peak addition and deletion, thus achieving a refined parameter estimation set via automated computation followed by statistical confirmation.

Software is available from the author (M.L.J.) upon written request.

Acknowledgments

The authors acknowledge the support of the National Science Foundation Science and Technology Center for Biological Timing at the University of Virginia (NSF DIR-8920162), the General Clinical Research Center at the University of Virginia (NIH RR-00847), and the University of Maryland at Baltimore Center for Fluorescence Spectroscopy (NIH RR-08119 and NIH R01 AG14799).

[5] Modeling of Oscillations in Endocrine Networks with Feedback

By Leon S. Farhy

General Principles in Endocrine Network Modeling

Numerous studies document that the hormone delivery pattern to target organs is crucial to the effectiveness of their action. Hormone release could be altered by pathophysiology and differences in endocrine output mediate important intraspecies distinctions, as, for example, some of the sexual dimorphism in body growth and gene expression in humans and rodents. Accordingly, the mechanisms controlling the dynamics of various hormones had lately become the object of extensive biomedical research. Intuitive reconstruction of endocrine axes is challenged by their high complexity, due to multiple intervening time-delayed nonlinear feedback and feedforward inputs from various hormones and/or neuroregulators. Consequently, quantitative methods have been developed to complement qualitative analysis and laboratory experiments and reveal the specifics of hormone release control. The emerging mathematical models interpret endocrine networks as dynamic systems and attempt to simulate and explain their temporal behavior.[1–6]

This chapter focuses on the mathematical approximation of endocrine oscillations in the framework of a modeling process structured in three formal phases:

1. Data analysis (examining the available data). We start by studying the available observations and experimental results, by examining the hormone time series, and determining the specifics of the observed profiles. This might include pulse detection, analysis of the variability and orderliness, verifying the baseline secretion and half-life, and detecting the frequency of the oscillations. We identify those phenomena that should be explained by the modeling effort, for example, some specific property of the hormone profiles, combined with selected feedback experiments.

2. Qualitative analysis (designing the formal network). This stage uses the information collected in phase 1 and outlines an intuitive functional scheme of the systems underlying physiology. Qualitative analysis of the available data[7] identifies the key elements and their interaction and organizes them as a set of nodes and conduits in a *formal endocrine network*. The *main hypothesis* states that this formal network explains the selected in phase 1 specifics in the experimental data.

3. Quantitative analysis (dynamic modeling). At this phase the endocrine network is interpreted as a dynamic system and described with a set of coupled ordinary differential equations (ODE). They give the time derivative of each network node and approximate all system positive and negative dose–responsive control loops. The parameters in the ODEs must have a clear physiological meaning and are determined by comparing the model output with the available data (phase 1) as we attempt to address the main hypothesis (phase 2).

[1] L. S. Farhy, M. Straume, M. L. Johnson, B. P. Kovatchev, and J. D. Veldhuis, *Am. J. Physiol. Reg. Integr. Comp. Physiol.* **281,** R38 (2001).

[2] L. S. Farhy, M. Straume, M. L. Johnson, B. P. Kovatchev, and J. D. Veldhuis, *Am. J. Physiol. Reg. Integr. Comp. Physiol.* **282,** R753 (2002).

[3] D. M. Keenan and J. D. Veldhuis, *Am. J. Physiol.* **281,** R1917 (2001).

[4] D. M. Keenan and J. D. Veldhuis, *Am. J. Physiol.* **280,** R1755 (2001).

[5] L. Chen, J. D. Veldhuis, M. L. Johnson, and M. Straume, *in* "Methods in Neurosciences," p. 270. Academic Press, New York, 1995.

[6] C. Wagner, S. R. Caplan, and G. S. Tannenbaum, *Am. J. Physiol.* **275,** E1046 (1998).

[7] O. Friesen and G. Block, *Am. J. Physiol.* **246,** R847 (1984).

The outcome of the modeling effort is a *conditional* answer to the main hypothesis. It formulates necessary physiological assumptions (additional to the main hypothesis) that would allow the formal network to explain the observed data specifics. This further refines the hypothesis and generates new questions to be addressed experimentally.

The general modeling scheme anticipates that the qualitative analysis of the hormone secretion dynamics outlines the formal endocrine network by determining its nodes and conduits. As previously discussed,[7] the main source of oscillations in biology is feedback loops with delay. However, not every network with feedback generates periodic behavior.[8] The main goal of this work is to illustrate via a series of abstract examples different conditions under which oscillations can emerge. To this end, we perform quantitative analysis on various abstract endocrine networks, interpreted as dynamic systems. Thus, we will be mainly concerned with phase 3 (previous) and its relations to phases 1 and 2.

We start by describing the approximation of the basic element of an endocrine network: the dynamics of the concentration of a single hormone, eventually controlled by one or more other regulators (system nodes). Further, this is used in the simulation and analysis of different feedback networks. The main concepts are illustrated on abstract 2-node/1-feedback reference models. System parameters are introduced on the basis of their physiological meaning and the effect of their modification is examined. Oscillations due to perturbations of systems with damped periodicity are distinguished from oscillations of systems with a true periodic solution (limit cycle). Additionally, we simulate basic laboratory experimental techniques, discuss some of their limitations, and suggest alternatives to reveal more network details.

It should be noted that the theory behind most of the examples in this chapter is not trivial. This is especially valid for those models that include one or more direct delays in the core system. We avoid the abstract mathematical details to make the presentation accessible to a variety of bioscientists. The simulated networks are abstract and do not correspond to a particular endocrine system. However, the constructs and the modeling techniques can be easily adapted to fit a particular physiology.

Simulating the Concentration Dynamics of a Single Hormone

In this section we describe the quantitative approximation of the concentration dynamics of a single hormone in an abstract pool, where it is secreted (not synthesized). As described elsewhere,[9] we assume that the

[8] R. Thomas, R. D'Ari, and N. Thomas, *in* "Biological Feedback." CRC Press, Boca Raton, FL, 1990.

hormone concentration rate of change depends on two processes—secretion and ongoing elimination. The quantitative description is given by the ordinary differential equation

$$\frac{dC}{dt} = -\alpha C(t) + S(t) \tag{1}$$

Here, $C(t)$ is the hormone concentration in the corresponding pool, t is the time, $S(t)$ is the rate of secretion, and the elimination is supposed to be proportional to the concentration.

Deconvolution technique, employed to describe hormone pulsatility,[9] can be used as an alternative approach to introducing Eq. (1). In this context, the observed hormone concentration is described by a convolution integral

$$C(t) = \int_0^t S(\tau)E(t-\tau)d\tau \tag{2}$$

where S is a secretion function and E describes the removal of the hormone from the pool. For the purposes of this presentation, E is required to correspond to a model with one half-life. In particular, we assume that the elimination function $E(t)$ satisfies the initial value problem

$$\frac{dE(t)}{dt} = -\alpha E(t)$$
$$E(0) = 1 \tag{3}$$

with some rate of elimination $\alpha > 0$. Consequently, it is easy to see that Eqs. (2) and (3) imply that the right-hand side of Eq. (1) describes the rate of change of $C(t)$. And since the solution of Eq. (3) is the function $E(t) = e^{-\alpha t}$ the hormone concentration [the solution of Eq. (1)] is described as the convolution integral

$$C(t) = \int_0^t S(\tau)e^{-\alpha(t-\tau)}d\tau$$

Now, suppose that the secretion rate $S = S_A$ (of a hormone A) does not depend explicitly on t and is controlled by some other hormone B. We write $S_A = S_A[C_B(t)]$, where $C_B(t)$ is the concentration of B. In the sequel, S_A is called a control function and its choice, albeit arbitrary to some extent, should conform to a set of general rules.

1. Minimal and maximal exogenous levels: Denote by $C_{A,\min}$ and by $C_{A,\max}$ the minimal and maximal values (experimentally established or hypothetical) for the concentration of hormone A. Typically (but

[9] J. D. Veldhuis and M. L. Johnson, *Methods Enzymol.* **210,** 539 (1992).

not always), $C_{A,\min}$ is associated with the baseline secretion and $C_{A,\max}$ corresponds to the maximal attainable concentration of exogenous A (on a variety of conditions, including responses to external submaximal stimulation). Accordingly, the control function S_A must satisfy the inequalities

$$C_{A,\min}/\alpha \leq \min(S_A) \leq \max(S_A) \leq C_{A,\max}/\alpha$$

2. Monotonous and nonnegative: The control function must be nonnegative, since the secretion rate is always nonnegative, and monotone (with some rare exceptions briefly mentioned in the sequel). It will be monotone increasing if it represents a positive control. If the control is negative, it will be decreasing.

There are many ways to introduce a control function in an acceptable mathematical form. As many authors do, we use nonlinear, sigmoid functions, known as up- and down-regulatory Hill functions[8]:

$$F_{\text{up(down)}}(G) = \begin{cases} \dfrac{[G/T]^n}{[G/T]^n + 1} & \text{(up) or} \\[2ex] \dfrac{1}{[G/T]^n + 1} & \text{(down)} \end{cases} \qquad (4)$$

where $T > 0$ is called a threshold and $n \geq 1$ is called a Hill coefficient. It should be noted that $F_{\text{up}} = 1 - F_{\text{down}}$ and $F_{\text{up(down)}}(T) = 1/2$. These functions are exemplified in the plots in Fig. 1 (for $n = 5$ and $T = 50$). They are monotone and map $F: (0, \infty) \to (0, \infty)$; the Hill coefficient n controls the slope (which also depends on T), and the inflection point I_F is given by

$$I_F = T\left(\frac{n-1}{n+1}\right)^{\frac{1}{n}} \qquad \text{for } n \geq 2$$

When $n = 1$ (Michaelis–Menten type equation) the function has no inflection point and its profile is a branch of a hyperbola. If n is large (values, as large as 100, exist in biology[10,11]) the control function acts almost as an on/off switch.

Using Hill functions, we write the term controlling the secretion of A in the form

$$S_A(C_B) = aF_{\text{up(down)}}(C_B) + S_{A,\text{basal}} \qquad (5)$$

where $S_{A,\text{basal}} \geq 0$ is independent of B and controls the basal secretion of A.

[10] P. V. Vrzheshch, O. V. Demina, S. I. Shram, and S. D. Varfolomeev, *FEBS Lett.* **351**(2), 168 (1994).

[11] T. Mikawa, R. Masui, and S. Kuramitsu, *J. Biochem.* **123**(3), 450 (1998).

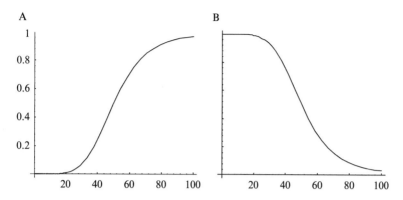

FIG. 1. Exemplary profiles of up-regulatory (A) and down-regulatory (B) Hill functions. In both examples $n = 5$ and $T = 50$.

The quantities $(a + S_{A,\text{basal}})/\alpha$ and $S_{A,\text{basal}}/\alpha$ represent the previously-mentioned $C_{A,\text{max}}$ and $C_{A,\text{min}}$, respectively.

As mentioned earlier, on certain occasions, the monotonousness of the control function may be violated. For example, if might happen that at low to medium concentrations a substance is a stimulator, while at high concentrations it is an inhibitor. Thus, the control function is nonmonotonous and can be written as a combination of Hill functions[8]:

$$S_A(G) = a\frac{[G/T_1]^{n_1}}{[G/T_1]^{n_1} + 1}\frac{1}{[G/T_2]^{n_2} + 1}, \qquad T_1 < T_2$$

Next, assume that instead of one, two hormones control the secretion of A. We denote them by B and C with corresponding concentrations $C_B(t)$ and $C_C(t)$. The control function $S_A = S_A(C_B, C_C)$ depends on the specific interaction between A from one side, and B and C from another.[8] For example, if both B and C stimulate the secretion of A

$$S_A(C_B, C_C) = a_B F_{\text{up}}(C_B) + a_C\, F_{\text{up}}(C_C) + S_{A,\text{basal}} \qquad (6)$$

if B and C act independently, or

$$S_A(C_B, C_C) = a F_{\text{up}}(C_B) F_{\text{up}}(C_C) + S_{A,\text{basal}} \qquad (7)$$

if B and C act simultaneously (the secretion of A requires the presence of both). On the other side, if, for example, the secretion of A is stimulated by B, but suppressed by C, the control function can be introduced as

$$S_A(C_B, C_C) = a F_{\text{up}}(C_B) F_{\text{down}}(C_C) + S_{A,\text{basal}} \qquad (8)$$

or

$$S_A(C_B, C_C) = a_B F_{\text{up}}(C_B) + a_C F_{\text{down}}(C_C) + S_{A,\text{basal}} \qquad (9)$$

Note, that Eq. (8) simulates a noncompetitive and simultaneous action of B and C. If B and C compete as they control the secretion of A, the secretion term can be described with a modified Hill function:

$$S_A(C_B, C_C) = a \frac{(C_B/T_B)^{n_B}}{(C_B/T_B)^{n_B} + (C_C/T_C)^{n_C} + 1} + S_{A,\text{basal}} \qquad (10)$$

Oscillations Driven by a Single System Feedback Loop

In this section we discuss in detail networks with a single (delayed) feedback loop that can generate oscillatory behavior. We focus on 2-node/1-feedback networks, in which the concentration of one hormone A regulates the secretion of another hormone B, which in turn controls the release of A. This construct can generate oscillations, even if there is no explicit (direct) delay in the feedback.* However, in this case the oscillations will fade to the steady state of the system. A nonzero delay and a large nonlinearity in the control functions (sufficiently high Hill coefficients) guarantee steady periodic behavior, due to the existence of a nontrivial limit cycle. On the other hand, a network may incorporate a single feedback loop by means of only one or more than two nodes. We comment on some peculiarities of such models in the last section.

Formal 2-node/1-Feedback Network

We study the abstract endocrine networks shown in Fig. 2. These particular examples anticipate that two hormones, A and B, are continuously secreted (driven by nonrhythmic excitatory input) in certain pool(s) (systemic circulation, portal blood, etc.), where they are subject to elimination. The release of hormone B is up-(down-)regulated by hormone A. Hormone B itself exerts a negative (positive) delayed feedback on the secretion of A. The A/B interactions are assumed to be dose responsive. The resulting delayed control loop is capable of driving hormone oscillations, if certain conditions (discussed later) are provided.

To formalize the networks depicted in Fig. 2, we denote the concentrations of hormones A and B by $C_A(t)$ and $C_B(t)$, respectively. We assume that the elimination of each hormone is proportional to its concentration with positive constants α and β. The secretion rate S_A of A is supposed to depend on the history of the concentration of B and vice versa. In particular, we assume that $S_A(t) = S_A\{h_1[C_B(t)]\}$ and $S_B(t) = S_B\{h_2[C_A(t)]\}$.

---

* The thresholds in the control functions provide implicit delays in the corresponding conduits.

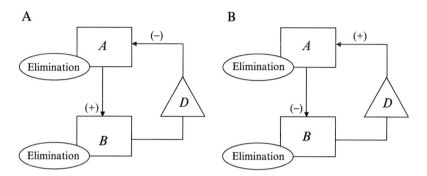

FIG. 2. Formal network of a two-node/one-feedback oscillator. (A) A network in which the main hormone B is stimulated; (B) a model in which B is inhibited. D denotes a delay in the interconnection. In both networks A and B are subject to elimination.

The functional h_1 (h_2) incorporates the lag in the action of B on A (A on B). To formally introduce the delays, one can account for the time-averaged effect of the hormone action in a past time interval related to the current moment.[4] However, this method requires two parameters for each delayed action—the onset and the termination of the delayed action.[4] Here, to keep the model as minimal as possible, we use a "direct" delay (with only one parameter for each delayed control action) and assume that the secretion control functions can be written as

$$S_A(t) = S_A[C_B(t - D_B)] \qquad \text{and} \qquad S_B(t) = S_B[C_A(t - D_A)]$$

with some nonnegative delay times D_A and D_B. Then, the system of ordinary (nonlinear) delayed differential equations, which describes a formal two-node/one-feedback endocrine network (Fig. 2), has the form

$$\frac{dC_A}{dt} = -\alpha C_A(t) + S_A[C_B(t - D_B)]$$

$$\frac{dC_B}{dt} = -\beta C_B(t) + S_B[C_A(t - D_A)] \tag{11}$$

with some elimination constants α, $\beta > 0$, lag times D_A, $D_B \geq 0$, and secretion rate control functions S_A, $S_B \geq 0$.

Reference Systems

To describe the dose-responsive relationships between A and B, corresponding to the network from Fig. 2A, we use the recommendations outlined in "Hormone release approximation" [Eq. (5)]. We write the control functions that appear in (11) as follows:

$$S_A[C_B(t - D_B)] = aF_{\text{down}}[C_B(t - D_B)] + S_{A,\text{basal}}$$

$$S_B[C_A(t - D_A)] = bF_{\text{up}}[C_A(t - D_A)] + S_{B,\text{basal}}$$

With this special choice, the core system of first-order nonlinear differential equations, describing the network from Fig. 2A, have the form

$$\frac{dC_A}{dt} = -\alpha C_A(t) + S_{A,\text{basal}} + a \frac{1}{[C_B(t - D_B)/T_B]^{n_B} + 1}$$

$$\frac{dC_B}{dt} = -\beta C_B(t) + S_{B,\text{basal}} + b \frac{[C_A(t - D_A)/T_A]^{n_A}}{[C_A(t - D_A)/T_A]^{n_A} + 1} \tag{12}$$

The units in this model are as follows:

C_A, C_B, T_A, T_B	mass/volume
$a,b,S_{A,\text{basal}}$, $S_{B,\text{basal}}$	mass/volume/time
α,β	time^{-1}
D_A, D_B	time

However, in the sequel we avoid specifying the specific unit and the simulated profiles have arbitrary magnitude, which could be rescaled with ease to fit a desired physiology.

In most of the simulations we assume no basal secretions and a direct action of A on B (no delay). This transforms the core equations [Eq. (12)] into

$$\frac{dC_A}{dt} = -\alpha C_A(t) + a \frac{1}{[C_B(t - D_B)/T_B]^{n_B} + 1}$$

$$\frac{dC_B}{dt} = -\beta C_B(t) + b \frac{[C_A(t)/T_A]^{n_A}}{[C_A(t)/T_A]^{n_A} + 1} \tag{13}$$

Note, that solving these equations for $t \geq t_0$ requires the initial condition for C_B to be given on the entire interval $[t_0 - D_B, t_0]$.

From the special form of Eq. (13) we could easily derive that after some time (depending on the initial conditions), the solutions will be bounded away from zero and from above. More formally, for any $\varepsilon > 0$ (and we may choose ε as small as we like), there exists $t_0 > 0$ (depending on ε, the initial conditions and the system parameters), such that for $t > t_0$ the following inequalities hold and provide upper and lower bounds on the solution of Eq. (13):

$$0 < \frac{a}{\alpha} \frac{1}{(b/[\beta T_B])^{n_B} + 1} - \varepsilon \leq C_A(t) \leq \frac{a}{\alpha} + \varepsilon$$

$$0 < \frac{b}{\beta} \frac{1}{(T_A/\min C_A)^{n_A} + 1} - \varepsilon \leq C_B(t) \leq b/\beta + \varepsilon \tag{14}$$

The upper bounds above are absolute system limits. For example, the model response to exogenous A-bolus cannot exceed the value b/β. However, since $C_A < a/\alpha$, we get from Eq. (14) that the actual endogenous peak concentration of B will never reach b/β. In fact, if there is no external input of energy in the system, it will be less than

$$C_B(t) \leq \frac{b}{\beta} \frac{1}{(\alpha T_A/a)^{n_A} + 1} < \frac{b}{\beta} \tag{15}$$

Hence, changes in four parameters (a, α, n_A, T_A) can model a difference between the maximal amplitude of the internally generated peaks and the eventual response to external stimulation. All estimates may be refined through a recurrent procedure inherent in the core system [Eq. (13)]. For example, one can combine the two inequalities Eq. (14) to get an explicit lower bound for C_B:

$$\frac{b}{\beta} \frac{1}{\left\{ \dfrac{\alpha T_A[(b/[\beta T_B])^{n_B} + 1]}{a} \right\}^{n_A} + 1} \leq C_B(t) \tag{16}$$

Accordingly, we can use this to write an explicit upper bound for C_A:

$$C_A \leq \frac{a}{\alpha} \frac{1}{\left(\dfrac{C_{B,\min}}{T_B} \right)^{n_B} + 1} \leq \frac{a}{\alpha} \frac{1}{\left(\dfrac{M}{T_B} \right)^{n_B} + 1},$$

$$\text{where } M = \frac{b}{\beta} \frac{1}{\left\{ \dfrac{\alpha T_A\left[\left(\dfrac{b}{\beta T_B} \right)^{n_B} + 1 \right]}{a} \right\}^{n_A} + 1}$$

These inequalities can help to determine reasonable values for the model parameters.

It is easy to see that (since the control functions are monotonously decreasing and increasing) the system Eq. (13) has a unique fixed point (steady state). It can be shown that if there is no delay $(D_A = D_B = 0)$ the fixed point is asymptotically stable (a node or a focus) and attracts all trajectories in the phase space (Fig. 3A). However, even a single nonzero delay [as in Eq. (13)] might change the properties of the steady state. The particular stability analysis is nontrivial, and consists of investigating the real part of eigenvalues, which are roots of equation containing a transcendental term, involving the delay. In the examples that follow, we will encounter one of the two situations depicted in Fig. 3: the steady state will be either an attracting focus (Fig. 3A) or a repellor (Fig. 3B), and in the latter case there

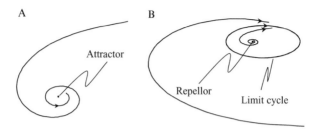

FIG. 3. Illustrative trajectories in the space (C_A, C_B) if the steady state is an attractor (A) or a repellor (B). In the latter case, a unique asymptotically stable periodic solution acts as a limit cycle and attracts all other trajectories (except the fixed point).

will exist a unique asymptotically stable periodic solution (which encircles the fixed point in the phase space) acting as a global limit cycle by attracting all trajectories (except the one originating from the fixed point).

Oscillations Generated by a Periodic Solution

In this section we present two specific examples describing the networks in Fig. 2. The core system of delayed ODE for the reference models will have unique periodic solution and unique repelling fixed point (Fig. 3B).

Consider a construct, described by the following core equations:

$$\frac{dC_A}{dt} = -1C_A(t) + 5\frac{1}{[C_B(t-3)/20]^2 + 1}$$

$$\frac{dC_B}{dt} = -2C_B(t) + 500\frac{[C_A(t)/5]^2}{[C_A(t)/5]^2 + 1} \tag{17}$$

These equations simulate the network shown in Fig. 2A (*A* is a stimulator). The parameters were chosen to guarantee stable oscillations (Fig. 4). Later, we show how the parameter choice affects the periodicity.

Even in this simple example, we have a variety of possibilities to model the specific interactions between *A* and *B*. In the previous example we have surmised:

1. The maximal attainable amplitude of C_B is 250.
2. The maximal attainable amplitude of C_A is 5.
3. The threshold T_A is higher than the endogenous levels of C_A.
4. The threshold T_B is approximately 6-fold lower than the highest endogenous levels of C_B.

It follows from 2 and 3 that the response of *B* to endogenous stimulation is not full. However, a high exogenous bolus of *B* elicits dose-dependent

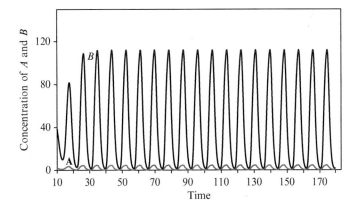

Fig. 4. Dynamics of the concentration of A (the lower profile) and B for the reference model described by Eq. (17).

release of B secretion at levels higher then the typical endogenous B concentration. It is easy to see that due to 2 the maximal endogenous B concentration is less than 125. Due to the choice of T_B (see 4), B almost fully suppresses the release of A between pulses, which in turn results in low intervolley B secretion.

To simulate the network from Fig. 2B (A is an inhibitor) we use the following reference system of delayed ODEs:

$$
\begin{aligned}
\frac{dC_A}{dt} &= -1C_A(t) + 50\frac{[C_B(t-3)/20]^2}{[C_B(t-3)/20]^2 + 1} \\
\frac{dC_B}{dt} &= -2C_B(t) + 500\frac{1}{[C_A(t)/5]^2 + 1}
\end{aligned}
\tag{18}
$$

The system parameter a in Eq. (17) was increased 10-fold [compared to Eq. (18)] to guarantee the existence of a periodic solution.

Simulation of Feedback Experiments

The success of a modeling effort is frequently measured by the capability of the construct to reproduce pivotal feedback experiments. Accordingly, we discuss the correct way of modeling and the system reaction to three common experimental techniques, aimed to disclose the specific linkages within an endocrine system.

Antibody Infusion. The introduction of an antibody (Ab) to a certain substance, referred here as S, is generally accompanied by a transformation of S, which results in effectively removing S from the system. The corresponding rate depends on the specific chemical reaction between Ab and S, and increasing the elimination constant of S (corresponding to the pool where Ab is administered) would model the removal. It remains possible that the reaction specifics change the single half-life pattern into a multiple half-life model. However, the single half-life approximation still might be sufficient in a variety of simulations.

To exemplify the idea we simulated variable removal of the inhibitor A in the reference model described by Eq. (18). Three simulations were performed, in which the coefficient β was increased 2-fold (left), 6-fold (middle), or 15-fold (right) at time $t = 75$.

The plots in Fig. 5A capture a very interesting phenomenon predicted by the model: the decrease in the peak amplitudes of B, even though an inhibitor is removed from the system. In the current model, this is explained by the actual increase of the rate at which A initiates its rise and reaches its action threshold, which, in turn, promotes an earlier suppression of B secretion.

Sensitivity Modification. Modifying the profiles of the control function models alterations in system sensitivity. For example, if the sensitivity of

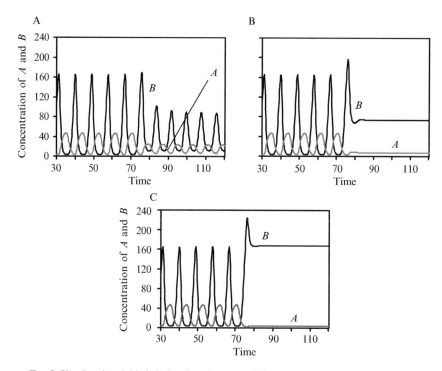

Fig. 5. Simulated variable infusion (starting at $t = 75$) of antibody to the inhibitor A in the reference model outlined in Eq. (18). The plots depict low (A), medium (B), or almost complete (C) removal of A.

certain cell group depends on the number of opened receptors, we could simulate receptor blockage/stimulation via changing the parameters of the corresponding control function. In the model described in Eq. (17), this would correspond to changes in the threshold, or in the Hill coefficient. Reducing (increasing) a threshold results in sensitivity increase (decrease). Changes in the Hill coefficient affect the slope of the control function. In general, increasing the Hill coefficient slightly changes the frequency and the amplitude, without affecting the pulsatility character of the profiles. In contrast, a decrease could effectively disturb the oscillations by preventing the system from overshooting the steady state.

We illustrate the effect of changing all thresholds and Hill coefficients in Eq. (17) (Fig. 6). An increase in n_B or n_A (Fig. 6A and C, left panels) produced a slight change in the frequency and amplitude. A decrease in n_B or n_A resulted in pulse shrinking (Fig. 6C, right panel) or in loss of periodicity (Fig. 6A, right panel) if the control functions can no longer provide the necessary inertia for overshooting the steady-state value. Increasing T_B from 20 to 80 (Fig. 6B, right panel) results in a condition in which B cannot exert the necessary suppression on A. The concentration of B is limited from above and increasing its action threshold gradually obliterates the effect of the delay containing term. Decreasing T_B to 0.2 has no visual effect on the simulated profiles (Fig. 6B, left panel). The pulsatility is not affected because the suppressive action of B on A is not modified. It starts only somewhat earlier, but there is still a 3-h delay in this action, which, in this particular model, is sufficient to maintain oscillations. The analysis of the effect produced by changes in T_A is somewhat different. Both increasing and decreasing might affect the oscillations. When T_A is decreased, even a small amount of A is sufficient to produce a full response, which obliterates the pulsatility (Fig. 6D, left panel). The fact that the concentration of A is bounded from below independently of T_A is crucial [Eq. (14)]. Increasing T_A results in a left shift of the control function S_B, thus, preventing A from stimulating B, which in turn reduces the oscillations (Fig. 6D, right panel).

A more formal approach to explaining the reduction in the range of the oscillations (the "shrinking" of the profile) would consist of (recursive) application of the inequalities [Eq. (14)]. For example, from the right-hand side of Eq. (14) it is evident that if $T_A \rightarrow 0$ then $C_B \rightarrow b/\beta$ and if $T_B \rightarrow \infty$ then $C_A \rightarrow a/\alpha$.

Exogenous Infusion. The correct way to simulate exogenous infusion of a hormone, which is also a system node, would be to add an infusion term to the right-hand side of the corresponding ODE. This term should correspond to the infusion rate profile in the real experiment. Mathematically, it might be interpreted as change in the basal secretion. In terms

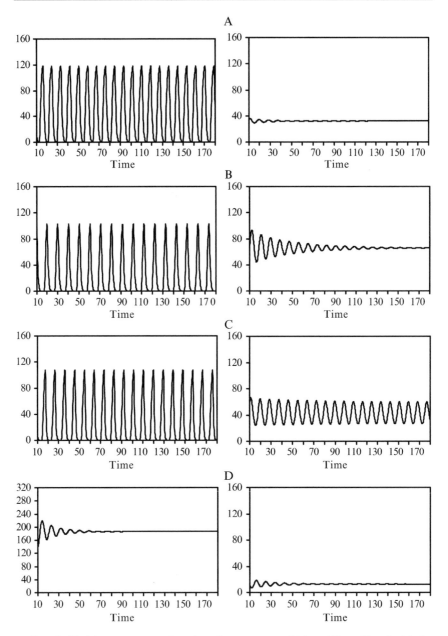

FIG. 6. Model response to alterations in system sensitivity. All profiles depict the dynamics of C_B (t). (A) Changing n_B from 2 to 10 (left) and to 1 (right); (B) changing T_B from 20 to 0.2 (left) and to 80 (right); (C) changing n_A from 2 to 20 (left) and to 2/3 (right); (D) changing T_A from 5 to 1/40 (left) and to 15 (right).

of the specific model described by Eq. (11), if we are simulating infusion of hormone B, the corresponding equation changes as follows:

$$\frac{dC_B}{dt} = -\beta C_B(t) + S_B[C_A(t - D_A)] + \inf(t) \tag{19}$$

where $\inf(t)$ is the infusion rate term. The solution of the previous equation is the sum of both endogenous and exogenous concentrations of B. To follow the distinction explicitly, a new equation should be added to the system:

$$\frac{dC_{\inf}}{dt} = -\beta C_{\inf}(t) + \inf(t)$$

and $C_B(t)$ has to be replaced by $C_B(t) + C_{\inf}(t)$ in all model equations, except the one that describes the rate of change of the concentration of B. To sum up, the core equations are

$$\frac{dC_A}{dt} = -\alpha C_A(t) + S_A\{[C_B + C_{\inf}](t - D_B)\}$$

$$\frac{dC_B}{dt} = -\beta C_B(t) + S_B[C_A(t - D_A)] \tag{20}$$

$$\frac{dC_{\inf}}{dt} = -\beta C_{\inf}(t) + \inf(t)$$

The model above [Eq. (20)] is in essence a 3-node/1-feedback construct, where exogenous B is the new node. A particular example, illustrating infusion simulation is shown later in this section (see "Identifying Nodes, Controlling the Oscillations").

Oscillations Generated by a Perturbation

In the reference models from the previous section the pulsatility was generated by a system that has a unique periodic solution and a unique fixed repelling point. The purpose of this section is to demonstrate that oscillations may occur as a result of disrupting a system that does not have a periodic solution, and its fixed point is an asymptotically stable focus (Fig. 3A).

We illustrate this concept on an earlier example. Figure 6B (right panel) depicts the profile of the solution to the following delayed ODE:

$$\frac{dC_A}{dt} = -1C_A(t) + 5\frac{1}{[C_B(t-3)/80]^2 + 1}$$

$$\frac{dC_B}{dt} = -2C_B(t) + 500\frac{[C_A(t)/5]^2}{[C_A(t)/5]^2 + 1} \tag{21}$$

The difference between this model and the reference construct [Eq. (17)] is in the 4-fold increase of the threshold T_B. In this case, there is no periodic solution and the unique fixed point attracts all trajectories in the phase space. Therefore, this system by itself cannot generate stable oscillations. However, if it is externally stimulated it can be removed from its steady state and oscillations will be detected. For example, assume that at $t = 350$ the secretion of B was briefly suppressed. This removes the trajectory in the phase space away from the fixed point and the system would have enough energy to initiate another waning pulse sequence (Fig. 7A). Moreover, if we allow for some periodic external control on the secretion, the hormone profile displays sustained pulsatility with bursts of variable amplitude (Fig. 7B). The frequency of the pulses is controlled by the coefficients of the core system [Eq. (21)], while the peak amplitudes follow the external stimulus.

If the perturbation is random, it generates pulses of approximately the same frequency as in the previous cases, but with highly variable amplitudes.

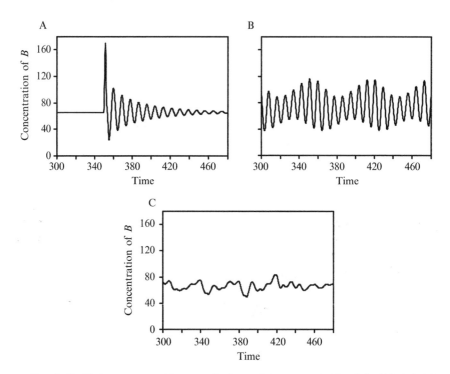

FIG. 7. Oscillations generated by perturbations of the system in Eq. (21). (A) A brief suppression of the secretion of B at $t = 350$. The rest of the profiles depict external periodic (B) or random (C) control on the coefficient b, which determines the release of B.

In the simulation presented in Fig. 7C we superimposed 40% Gaussian noise on the parameter b. Even though some peaks cannot be detected an overall pulse periodicity (the same as in Fig. 7A and 7B) is apparent.

In the previous examples, the perturbation was assumed to be external and independent of the core system. Later on, we show that a delayed system feedback could also provide enough energy and trigger oscillations in submodels with damped periodicity. In the three-node example from "Networks with Multiple Feedback Loops" a 2-node subsystem (with no direct delay in its feedback, and, therefore, without a periodic solution) is perturbed by a delayed system loop via the third node. This removes the whole system from its steady state and drives consecutive pulses during recurrent volleys.

Identifying Nodes, Controlling the Oscillations

When hormone A cannot be measured directly and is an inhibitor (the network in Fig. 2B) we can test whether it is involved in generating the oscillations of B by neutralizing the action (A-receptor blocker) or by removing (antibody infusion) A from its action pool. On the other hand, if A is a stimulator (Fig. 2A) a large constant infusion of A should remove the oscillations (by exceeding the action threshold, resulting in continuous full response from the target organ). This concept is exemplified in Fig. 8, which depicts two computer-generated predictions for the system response to exogenous infusion of hormone A [assuming that A stimulates B, Eq. (18)]. We simulated constant low (Fig. 8A) and high (Fig. 8B) infusion of A by increasing the basal A-secretion from zero to two different levels, starting at $t = 75$.

The model predicts gradual pulse "shrinking" toward the current steady-state level. If the exogenous administration of A is sufficiently high (Fig. 8B) the pulses wane and the secretion becomes constant. The profiles in Fig. 8 depict the numerical solution (concentration of hormone B) of the system

$$\frac{dC_A}{dt} = -C_A(t) + \text{Inf}(t) + 5\frac{1}{[C_B(t-3)/20]^2 + 1}$$
$$\frac{dC_B}{dt} = -2C_B(t) + 500\frac{[C_A(t)/5]^2}{[C_A(t)/5]^2 + 1} \tag{22}$$

with two different continuous infusion terms satisfying

$$\text{Inf}(t) = \begin{cases} 0 & \text{if } t \leq 75 \\ 1 \text{ or } 2 & \text{if } t \geq 76 \end{cases}$$

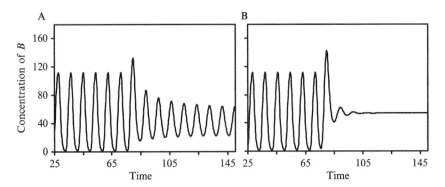

FIG. 8. System response [Eq. (22)] to exogenous infusion of A. The plots show simulation of constant low (A) and high (B) infusion of A starting at $t = 75$.

The parameters and control functions were chosen arbitrarily to simulate a network like the one in Fig. 2A, which generates stable oscillations.

Almost identical results (Fig. 5) can be achieved by simulating partial or complete removal of A in the case when A is an inhibitor (the network from Fig. 2B). This should be done by increasing the rate of elimination of A to simulate additional removal due to infusion of antibody (see "Simulation of Feedback Experiments" for details).

However, these experiments cannot disclose whether A is actually involved in a feedback with B, or acts merely as a trigger to remove a certain subsystem from its steady state. For example, consider the two networks shown in Fig. 9 and suppose that only the concentrations of hormone B can be measured.

Assume that E stimulates B, and its removal obliterates the secretion of B. Since E cannot be measured, we have no direct means to establish whether E is involved in a delayed feedback loop with B. Moreover, in both networks, constant high infusion of E (as proposed previously) removes the pulsatility and elicits constant secretion of B. Therefore, a more sophisticated experiment is required to reveal whether E is indeed involved in a feedback loop with B (Fig. 9A) or acts by perturbing the A–B subsystem (Fig. 9B). A possible approach would include blocking the exogenous E secretion with subsequent introduction of a single endogenous E bolus. The system response would be a single spike of B secretion, if the network were that, depicted in Fig. 9A, or a waning train of several B pulses if the network is the one, shown in Fig. 9B. Most importantly, the required suppression of exogenous E release must be achieved without affecting the putative A–B relationship.

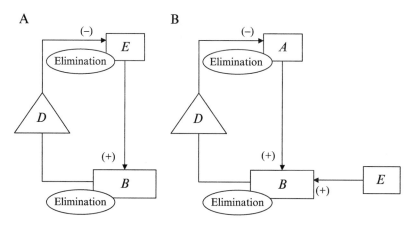

FIG. 9. Two hypothetical networks, in which a hormone E stimulates the secretion of B. E is either involved in a delayed feedback (A) or removes the subsystem A-B (B) from its steady state.

Separating Synthesis from Secretion

In certain cases, it would be appropriate to separate on a network level the hormone synthesis from its release. This would be important if a certain compound differently affects these processes. For example, let us consider again the network from Fig. 2A in an attempt to explain a rebound release of B following a withdrawal of continuous infusion of certain substance C. Assume that during the infusion of C the release of B was suppressed and that we have evidence that C is not affecting the release of A. A possible explanation of the rebound phenomenon would be that C affects the release of B, but not its synthesis. However, since all conduits in the network are affected in this experiment, the intuitive reconstruction of all processes involved is not trivial. The simulation requires introduction of a "storage" pool in which B is synthesized and packed for release and another pool (e.g., circulation) in which B is secreted. This adds a new equation to the model, describing the dynamics of the concentration of B in the storage pool. The following assumptions would be appropriate:

1. The concentration of B in the storage pool (P_B) is positively affected by the synthesis and negatively affected by the release.
2. The concentration P_B exerts a negative feedback on the synthesis of B and cannot exceed a certain limit P_{max}.
3. The rate of release of B from the storage pool is stimulated by the storage pool concentration but might be inhibited by the concentration of B in the exterior.

4. B is subjected to elimination only after it is secreted.

To provide an abstract example, assume that in the network from Fig. 2A we have in addition to A and B a new substance C that inhibits the secretion (competing with A), but does not affect the synthesis of B (Fig. 10).

Using Eq. (10) as a suitable form for the "competitive" control function, we can describe the network by the following system of delayed ODEs:

$$\frac{dC_A}{dt} = -\alpha C_A(t) + a \frac{1}{[C_B(t - D_B)/T_B]^{n_B} + 1}$$

$$\frac{dC_B}{dt} = -\beta C_B(t) + b \frac{[C_A(t)/T_{A,1}]^{n_{A,1}}}{[C_A(t)/T_{A,1}]^{n_{A,1}} + [C_C(t)/T_C]^{n_C} + 1} \frac{[P_B(t)/T_P]^{n_P}}{[P_B(t)/T_P]^{n_P} + 1}$$

$$\frac{dP_B}{dt} = c(P_{\max} - P_B) \frac{[C_A(t)/T_{A,2}]^{n_{A,2}}}{[C_A(t)/T_{A,2}]^{n_{A,2}} + 1}$$

$$- b\theta \frac{[C_A(t)/T_{A,1}]^{n_{A,1}}}{[C_A(t)/T_{A,1}]^{n_{A,1}} + [C_C(t)/T_C]^{n_C} + 1} \frac{[P_B(t)/T_P]^{n_P}}{[P_B(t)/T_P]^{n_P} + 1}$$

(23)

Here, for simplicity, we assumed that circulating B levels do not feed back on the secretion. This would correspond to a model with a much higher concentration in the storage pool than in the circulation. In the previous

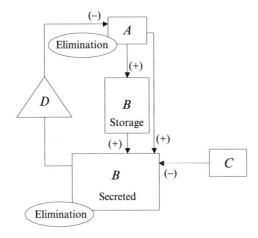

FIG. 10. Formal network depicting the system distinction between synthesis and release. C suppresses the release of B, but not its synthesis.

presentation c controls the rate of A-stimulated synthesis of B. The parameter θ represents the ratio between the volumes of the storage pool and the pool in which B is secreted. Typically, the second pool is larger and $\theta > 1$. We have supposed that the control functions, which correspond to the A-driven synthesis and release, are different with distinct thresholds $T_{A,1}$ and $T_{A,2}$, and corresponding Hill coefficients $n_{A,1}$ and $n_{A,2}$. The control, exerted on the secretion by the current concentration of B in the storage pool, is presented by the up-regulatory function $[P_B(t)/T_P]^{n_P}/\{[P_B(t)/T_P]^{n_P} + 1\}\}$. The following values were assigned to the parameters that appear in Eq. (23):

$$\alpha = 1; \quad \beta = 2; \quad \theta = 6; \quad a = 4; \quad b = 4000; \quad c = 2; \quad P_{max} = 1000;$$
$$T_{A,1} = 4; \quad T_{A,2} = 3; \quad T_B = 40; \quad T_C = 10; \quad T_P = 500;$$
$$n_{A,1} = 2; \quad n_{A,2} = 2; \quad n_B = 2; \quad n_C = 2; \quad n_P = 2;$$

The infusion term $C_C(t)$ is assumed to be a nonzero constant only during the time of infusion:

$$C_C(t) = \begin{cases} 0 & \text{if} \quad t < 55 \\ 500 & \text{if} \quad 56 < t < 95 \\ 0 & \text{if} \quad t > 96 \end{cases}$$

The model output is shown in Fig. 11 and the plots clearly demonstrate a B rebound following the withdrawal of C (Fig. 11A).

During the infusion the secretion of B is blocked, but not the synthesis and the concentration in the storage pool is elevated (Fig. 11B). The concentration of A increases (Fig. 11C), since low B levels cannot effectively block its release. Thus, the model explains the rebound jointly by the augmented concentration in the storage pool and the increased secretion of A.

Networks with Multiple Feedback Loops

The available experimental data might suggest that the release of a particular hormone B is controlled by multiple mechanisms, with different periodicity in the timing of their action. This implies that probably more than one (delayed) feedback loops regulate the secretion of B and the formal endocrine network may include more than two nodes. In determining the elements to be included in the core construct, it is important to keep track on the length of the delays in the feedback action of all nodes of interest. For example, if the goal were to explain events recurring every 1–3 h, the natural candidates to be included in the formal network would be

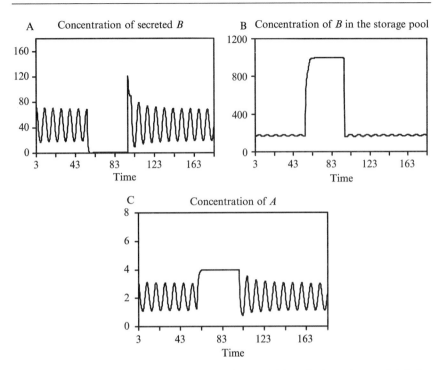

FIG. 11. Simulated rebound response following a withdrawal of continuous C infusion (timeline 55–95). (A) Concentration of secreted B (in the circulation). (B) Concentration of B in the storage pool. (C) A-concentration dynamics.

nodes, involved in feedback or feedforward relations with B with delays shorter than 3 h. Long feedback delays cannot account for high frequency events. In particular, if we hypothesize that a certain delayed feedback is responsible for a train of pulses in the hormone concentration profile, the direct delay must be shorter than the interpulse interval.

In this section we briefly discuss some features of abstract endocrine networks, incorporating more than one delayed feedback loop. Each loop accounts for its own oscillator mechanism and in what follows, we consider networks with two (delayed) feedback loops. Examples of 2-feedback constructs are shown in Fig. 12.

It should be noted that each of the two 3-node networks, shown in the middle panels of Fig. 12, could be reduced to its corresponding 2-node network from the top panels of Fig. 12. For example, let us consider the 3-node/2-feedback network shown in Fig. 12 (middle left panel). Assuming that both B and C can fully suppress the release of A, we can describe the formal network by the system of delayed ODE:

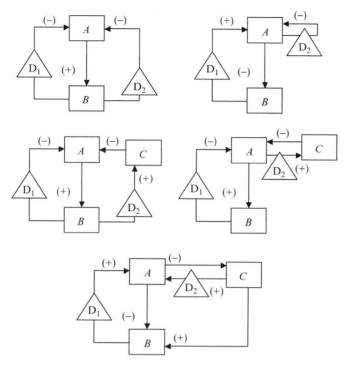

FIG. 12. Examples of hypothetical endocrine networks with more than one delayed feedback loops.

$$\frac{dC_A}{dt} = -3C_A(t) + 10{,}000 \frac{1}{[C_B(t)/100]^3 + 1} \frac{1}{[C_C(t)/70]^{20} + 1}$$

$$\frac{dC_B}{dt} = -2C_B(t) + 6000 \frac{[C_A(t)/500]^{40}}{[C_A(t)/500]^{40} + 1} \tag{24}$$

$$\frac{dC_C}{dt} = -3C_C(t) + 180 + 1320 \frac{[C_B(t-1.5)/200]}{[C_B(t-1.5)/200] + 1}$$

Here, for simplicity, we have assumed that there is no delay in the feedback $B \rightarrow A$. This system is capable of generating recurring multiphase volleys, by the mechanism described in "Oscillations Generated by a Perturbation" (Fig. 13).

However, analogous results can be achieved by reducing the 3-node network to a 2-node model with two feedbacks. In fact, the sequence of nodes and conduits $B \rightarrow C \rightarrow A \rightarrow B$ is, in essence, a negative 2-node delayed feedback loop: $B \rightarrow A \rightarrow B$. Therefore, it can be modeled in the

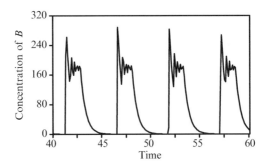

FIG. 13. Computer-generated output (concentration of B) of the core system Eq. (24).

usual way (by simply removing C from the system). The reduced network is
the one shown in Fig. 12 (upper left panel).

A corresponding simplified system of delayed ODEs could be

$$\frac{dC_A}{dt} = -3C_A(t) + 10{,}000 \frac{1}{[C_B(t)/100]^3 + 1} \frac{1}{[C_B(t - 1.5)/50]^3 + 1}$$

$$\frac{dC_B}{dt} = -2C_B(t) + 6000 \frac{[C_A(t)/500]^{40}}{[C_A(t)/500]^{400} + 1}$$

and the model output (not shown), even without any special efforts to adjust
the system parameters, is almost identical to the profile shown in Fig. 13.

Decreasing the number of equations from three to two reduces the
number of parameters to be determined and the time needed for solving the
equations numerically. This would be most important if multiple computer
runs are required. Therefore, adding the third node in the formal network
can be justified only if the goal is to simulate experiments involving C explicitly. And even then, the initial adjustment of the model would be significantly facilitated if C enters the system after the 2-node construct is
validated.

Note that if the network is more complex, the attempt to reduce the number
of nodes might not be beneficial. For example, the network shown in Fig. 12
(lower panel) cannot be transformed into a 2-node model, due to the high
system interconnectivity. We comment more on this in the next section.

Summary and Discussion

The mathematical methods presented in this chapter are tailored
to quantitatively interpret formal endocrine networks with (delayed)
feedbacks. The main goal is to illustrate different conditions, under which
oscillations can emerge.

The formal network itself consists of nodes and conduits, and is based on a qualitative analysis of available experimental data.[7] In our presentation the nodes are hormone concentrations in abstract pools, in which hormones are released or synthesized, under the control of other hormones. The conduits specify how the nodes interact within the network. The quantitative analysis of the formal network is based on approximation of the rate of change of a single system node. This essentially means that the dynamics of the hormone concentration is described with a single (delayed) ODE. To this end, we assume that the rate of change of hormone concentration depends on two processes—secretion and ongoing elimination. We work with a single half-life elimination model and express the control of the synthesis as a combination of sigmoid Hill functions, depending on the related nodes. The derivation of the ODE is demonstrated, along with a brief analysis of the properties of its solution to facilitate the actual determination of all system parameters.

The formal network is then interpreted as a dynamic system by combining all ODEs that describe system nodes dynamics. We exemplify the ideas on a 2-node/1-feedback model—one of the simplest meaningful examples of a network capable of generating and sustaining periodic behavior. In fact, a variety of systems display oscillatory behavior, driven by a single feedback loop. The simplest case is a 1-node/1-feedback network, in which a hormone after being secreted suppresses its own release, immediately or after some lag time. This system can generate periodic behavior only if the delay in the feedback is greater than zero. We do not discuss this case here.

A network may incorporate a single feedback loop in a more complex way, e.g., via a combination of two or more nodes. For example, simple stability analysis of the steady state shows that a 3-node/1-feedback network is capable of sustaining periodicity even without a delay in the feedback loop and relatively low Hill coefficients.[8,12] However, for a variety of practical cases, it is feasible to reduce the 3-node/1-feedback network to a 2-node/1-feedback construct as shown in the previous section.

Some specifics in endocrine network modeling are exemplified on two 2-node/1-feedback networks, in which the concentration of one hormone regulates the secretion of another, which in turn controls the release of the first hormone. This construct could generate oscillations even if there is no explicit delay in the feedback. However, it will be a damped periodicity, since the oscillations will fade and approach the steady state of the system. In contrast, a nonzero delay combined with a sufficiently large nonlinearity in the control functions (high Hill coefficients) guarantees steady periodic behavior, as all trajectories approach a nontrivial limit cycle.

[12] J. Richelle, *Bull. Cl. Sci. Acad. R. Belg.* **63**, 534 (1977).

We relate all parameters to their physiological meaning and analyze the solutions to our reference systems, which always have only one fixed point (steady state), which is either a repellor or an attractor (Fig. 3). In the first case the system has a unique limit cycle—a periodic solution, which attracts all trajectories in the phase space and, thereby, generates stable periodic behavior (Fig. 4). In the second case, the steady state is a focus and attracts all trajectories in the phase space. Therefore, the construct displays damped periodic behavior. In particular, if it is in a state close to the fixed point an external perturbation initiates a waning train of pulses (Fig. 7A). Therefore, oscillations might be generated even by a system that does not have a periodic solution, and its fixed point is an asymptotically stable focus. However, an external source of perturbations must be provided. Note that the frequency of the oscillations is largely independent of the external perturbation (Fig. 7).

We use the two reference systems to illustrate the modeling of three common experimental techniques: infusion of antibody to one of the nodes, sensitivity alterations, and exogenous infusion of one of the system hormones. We comment on the correct way to perform these approximations and examine the corresponding model response. In particular, the simulations illustrate conditions that might disrupt the periodicity.

Increasing the elimination rate of a hormone simulates infusion of antibody, and almost a complete removal of one of the nodes, and results in loss of periodicity (Fig. 5). Changes in the profiles of the control functions model alterations in system sensitivity. The analysis shows that if a model has a stable periodic behavior, the increase in one of the Hill coefficients would not change the system performance (Fig. 6A and C, left panels).[13] On the other side, a decrease in the same parameter may transform the steady state from a repellor into an attractor and affect the periodic behavior. Changes in the action thresholds may also affect the periodicity (Fig. 6B and D). Exogenous infusion can be simulated by a simple increase in the basal secretion, or by introducing a third node, in case we would like to distinguish between exogenous infusion and endogenous secretion of one and the same substance [Eqs. (19) and (20)].

We illustrate how these experiments may be used to disclose whether a certain hormone A is involved in generating the oscillations of another hormone B. The idea is to alter A in such way that the periodic B-profile is transformed into a constant nonzero secretion. When A inhibits B, we can neutralize its action (receptor blocker) or remove (antibody) it from the system. In the later case the model predicts that the periodicity disappears and is replaced by a stable B secretion (Fig. 8). Alternatively, if A stimulates B, a large continuous A infusion obliterates the oscillations by exceeding the action threshold and

[13] L. Glass and S. A. Kauffman, *J. Theor. Biol.* **39,** 103 (1973).

eliciting a unvarying full B response from the target organ (Fig. 5). Additionally, the model provides means to disclose whether A is actually involved in a feedback loop with B or generates oscillations by perturbing another subsystem (see "Identifying Nodes, Controlling the Oscillations").

To be able to capture a variety of feedback systems we separate on a network level the hormone synthesis from its release. The proper simulation requires a new "storage" pool in which the hormone is synthesized and stored, in addition to the pool, in which the hormone is secreted. We used this distinction to provide a plausible explanation of a rebound release, following withdrawal of an agent that suppresses the secretion, but not the synthesis.

We would like to emphasize the importance of keeping the model as minimal as possible while performing the initial qualitative analysis of the available experimental data. In general, formal endocrine networks might incorporate multiple feedbacks loops and nodes. However, long feedback delays cannot account for high-frequency events. Therefore, if the model attempts to explain pulses of a hormone that recur every H hours, it might be sufficient to include in the formal network only feedback loops with delay shorter than H. Moreover, if a feedback loop enters the network via a multiple-node subsystem, it might be possible to reduce the number of nodes and simplify the model without affecting its performance. The example provided in the previous section demonstrates a case in which we could safely remove a "passive" node from a feedback loop and still retain the overall periodic behavior.

Unfortunately, we cannot always reduce complex networks. The model shown in Fig. 12 (lower panel) is an example in which the system interconnectivity would not allow any simplification. Complex networks with intertwined feedback loops are considered elsewhere[2,3] and their analysis strongly depends on the specific physiology. It should be noted that in this chapter we do not consider more complicated cases, such as networks that have multiple steady states of different type, which is a significant complication. Such systems can be approached in the early stage of their analysis by Boolean formalization,[14,15] which serves as an intermediate between modeling phases 2 and 3 described in the first section. This method describes complex systems in simple terms and allows for preliminary finding of all stable and unstable steady states.

Acknowledgments

I acknowledge support by NIH Grant K25 HD01474 and I would like to thank my mentor, Dr. Johannes Veldhuis, for intellectual support and guidance.

[14] R. Thomas, *J. Theor. Biol.* **42,** 563 (1973).
[15] R. Thomas, *Adv. Chem. Phys.* **55,** 247 (1983).

[6] Measuring the Coupling of Hormone Concentration Time Series Using Polynomial Transfer Functions

By Christopher R. Fox, Leon S. Farhy, William S. Evans, and Michael L. Johnson

Introduction

Regulation of hormone release typically occurs via a complex endocrine network, utilizing feedforward and/or feedback relationships that couple system components. Multiple quantitative techniques have been developed to define different characteristics of a single hormone secretory process in such networks. For example, pulse detection algorithms and deconvolution analysis can characterize the secretory dynamics of a single hormone and describe its pulsatile nature.[1] However, less attention has been paid to appraising the eventual relationship between two or more simultaneously collected hormone concentration–time series. Several statistical techniques are available to relate the time course of the concentration of one hormone to that of a second, including cross-correlation, cross-approximate entropy,[2] and coincidence of peaks.[3] While each of these tools can roughly determine whether the secretions of two hormones are related, they cannot precisely predict the particular relationship between them. Therefore, approaches are needed that will allow evaluation of the coupling between different components of the endocrine network. A knowledge of the coupling will allow application of network modeling techniques[4] to predict the dynamics of a single hormone from other known components of the network, given the hypothesis that the release of each hormone can be determined using the concentrations of other system hormones and their biological half-lives. Additionally, knowing the nature of the coupling between two hormones will allow insights about whether alterations in this coupling may explain (at least in part) changes that occur in various pathophysiological states.

In this work, a method for estimating the strength of coupling and the relationship between two hormones is outlined. We report the performance of polynomial transfer functions as a tool to reconstruct the

[1] M. L. Johnson and M. Straume, *in* "Sex-Steroid Interactions with Growth Hormone," p. 318. Serona Symposia, 1999.

[2] S. M. Pincus and B. H. Singer, *Proc. Natl. Acad. Sci. USA* **93,** 2083 (1996).

[3] J. D. Veldhuis, M. L. Johnson, and E. Seneta, *J. Clin. Endocrinol. Metab.* **73,** 569 (1991).

[4] L. S. Farhy, *Methods Enzymol.* **384**(5), 2004 (this volume).

coupling between various combinations of hormone concentration–time series data derived from a known mathematical construct that describes certain hypothetical endocrine feedback networks. The use of synthetic data allows comparison between method-predicted coupling and known (model-defined) hormone interactions.

Methods

Definition of the Coupled System

Oscillating, periodic time-series data were created using a model consisting of three hormones (A, B, and C) linked by feedback and feedforward control loops. In this model, hormone A drives the secretion of hormone B. Hormone B exerts negative feedback to the release of A, and simultaneously stimulates (after a fixed time delay, D) the secretion of hormone C, which, in turn, suppresses the release of A. The schema for the coupling between A, B, and C, and the specific set of nonlinear differential equations, which describe this network, are shown in Fig. 1.[4] The output of this system with time steps of 0.015 (arbitrary time units) is depicted in Fig. 2.

Creation of Time-Series Hormone Data

The concentrations of hormones A, B, and C were reported at a fixed time interval of 0.015 to generate 6000 data points. To recreate examples of hormone time-series data, where hormone concentrations are measured repeatedly over a fixed time interval, but where there are significant time periods during which the hormone concentrations are not measured, time-series data for hormones A, B, and C were created by collecting every tenth data point beginning at an arbitrary point until 150 points had been sampled. Subsequently, a variable amount of noise (with SD equal to 1, 5, 10, 15, or 20% of the mean series value) was added. An example of the time-series data for hormone A with 1% versus 20% noise is shown in Fig. 3.

Analysis of Hormone Time-Series Data

One physiologically important mode of coupling is that the concentration of one, or more, hormones will stimulate, or inhibit, the secretion of another hormone. The time course of the concentration of a hormone in the serum can be described as a convolution integral of the time course of the secretion into the serum and the elimination kinetics for the removal

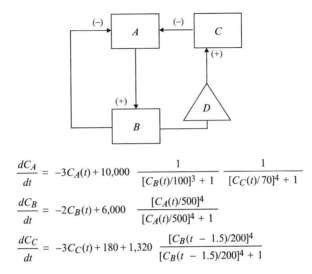

$$\frac{dC_A}{dt} = -3C_A(t) + 10,000 \ \frac{1}{[C_B(t)/100]^3 + 1} \ \frac{1}{[C_C(t)/70]^4 + 1}$$

$$\frac{dC_B}{dt} = -2C_B(t) + 6,000 \ \frac{[C_A(t)/500]^4}{[C_A(t)/500]^4 + 1}$$

$$\frac{dC_C}{dt} = -3C_C(t) + 180 + 1,320 \ \frac{[C_B(t - 1.5)/200]^4}{[C_B(t - 1.5)/200]^4 + 1}$$

FIG. 1. Top: Schema of the principal connections within a synthetic endocrine network that generates the hormone concentration–time series, used to test the coupling measure performance. Hormone A drives the secretion of hormone B, which in turn stimulates the secretion of hormone C [after a time delay (D)]. Hormones B and C suppress the release of hormone A. Bottom: Specific quantitative interpretation of the node-conduit relations (top panel) as a dynamic system, described by coupled nonlinear differential equations. The functions C_A, C_B, and C_C represent concentration of the hormone A, B, or C, respectively.

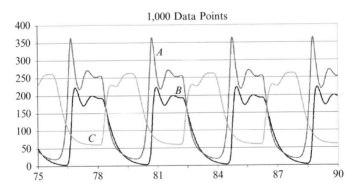

FIG. 2. Model-derived concentration profiles for hormones A, B, and C over 1000 data points.

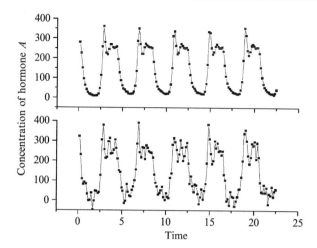

FIG. 3. The concentration of hormone A shown with the addition of random (measurement) noise with SD equal to 1% (top) and 20% (bottom) of the mean concentration of the hormone series.

of the hormone from the serum.[5] Thus, given the concentration–time series and the elimination kinetics, the secretion time series can be evaluated by a deconvolution procedure.[1]

To evaluate the coupling between the hormones the data were analyzed using the hypothesis "concentration of hormone X drives the secretion of hormone Y." (For simplicity, this is hereafter referred to as "X to Sec$Y$.") The secretion of hormones A, B, or C was calculated using waveform-independent deconvolution analysis.[1] The specified half-lives were set equal to ln2/(rate constant of elimination), as defined in the differential equations ($t_{1/2} = 0.231$, 0.347, and 0.231 for A, B, and C, respectively).

Linear, quadratic, and cubic polynomial transfer functions were evaluated by assuming that the secretion rate of hormone Y (Sec$Y$[time]) is a linear, quadratic, or cubic polynomial function of the time-shifted concentration of hormone X (X[time]):

Linear: $\text{Sec}Y[\text{time}] = \alpha + \beta\, X[\text{time} - \Delta t]$

Quadratic: $\text{Sec}Y[\text{time}] = \alpha + \beta\, X[\text{time} - \Delta t] + \gamma\, X[\text{time} - \Delta t]^2$

Cubic: $\text{Sec}Y[\text{time}] = \alpha + \beta\, X[\text{time} - \Delta t] + \gamma\, X[\text{time} - \Delta t]^2$
$\qquad\qquad\qquad + \delta\, X[\text{time} - \Delta t]^3$

[5] J. D. Veldhuis, M. L. Carlson, and M. L. Johnson, Proc. Natl. Acad. Sci. USA **84,** 686 (1987).

The parameters (α, β, γ, and δ) were evaluated by linear least-squares parameter estimation as a series of different lag times (Δt).

The main outcome is the percentage of the overall variance of the secretion of hormone Y that is accounted for by the concentration of hormone X transformed using the derived transfer function. It depends on the choice of the lag time (Δt) and its maximal value will be further referred to as strength of coupling for the relation "X to Sec$Y$."

The secretion control functions (Fig. 1) utilized to generate the synthetic data combine variations of the Hill equation, specifically,

$$(X[\text{time}]/K)^n/[1 + (X[\text{time}]/K)^n] \quad \text{(for upregulation)}$$

or

$$1/[1 + (X[\text{time}]/K)^n] \quad \text{(for downregulation)},$$

where K is the half-maximal effective concentration, and n is the Hill coefficient, that determines the steepness of the response.[4] This form of transfer function was chosen for the simulations because it is typical of the type of saturable nonlinear processes that are actually observed in biological systems. This Hill formulation could have been used for this analysis. However, several reasons justify the implementation of polynomials, instead of Hill functions, as surrogates for the infinite number of possible physiologically relevant forms.

First, in a real experimental situation the exact form of the transfer function will not be known a priori and it is possible that it cannot be described with a single Hill equation. In this regard, the polynomials have the advantage over Hill functions in that they are more flexible and capable of describing a larger variety of interactions. Second, for any given set of experimental, or simulated, data the transfer function is commonly not observed on its full domain of definition, and that would challenge the involved numerical methods. For example, the range of the concentration of A series in Fig. 2 is approximately between 0 and 350, while the half-maximal effective concentration of the Hill transfer function of the second differential equation is 500. Thus, the transfer function in this simulation is observed only over a small part of its domain of definition, below the half-maximal effective concentration. The Hill, and other, transfer functions over such a limited range can easily be described by a low-order polynomial. Lastly, the polynomial forms also have the advantage to be linear in the parameters while the Hill, and most other, possible formulations are nonlinear. The numerical analysis of linear forms is computationally much easier and significantly faster.

Results

Strength of Coupling for Each Combination of Hormones

The percent of the variance of the secretion of hormone *Y*, as calculated by waveform-independent deconvolution analysis, accounted for by the concentration of hormone *X* transformed using the transfer function was calculated for each possible hormone pair (i.e., concentration of *A* to the secretion of *B*, *B* to the secretion of *A*, *A* to the secretion of *C*, ...). The maximum percent of the variance for each pair, and the offset (measured in units, equal to 0.15 of the accepted model time unit) at which the maximum occurred, is summarized in Table I. Figures 4 and 5 show the percent of variance accounted for at each offset value up to a maximal offset of 12. In general, stronger coupling was observed when there was a direct link between the hormones (without any intervening signals) and when there was only one feedforward–feedback loop involved. Increasing the order of the polynomial transfer function generally improved the strength of coupling.

Optimal observed time lags shown in Table I are exactly as expected for data simulated according to the network as shown in Fig. 1. It is particularly interesting to note that the network contains two pathways for coupling from *B* to *A*. The direct one, *B* → *A*, with a lag of zero and a delayed one, *B* → *C* → *A*, with a delay of 10 time units. Figure 5 presents the analysis of the simulated data and it indicates two modes, one with 0 lag and about 20% of the variance and a second with a lag of 11 and about 70% of the variance. Thus, the dominate coupling pathway for this example is *B* → *C* → *A*.

TABLE I
Strength of Coupling[a]

Order of transfer function	A to SecB[b]	B to SecA	A to SecC	C to SecA	B to SecC	C to SecB
Offset[c]	0	11	12	0	10	0
Linear	73	61	91	62	90	52
Quadratic	84	62	93	61	94	54
Cubic	84	69	95	68	94	54

[a] Strength of coupling as the percentage of the variance of the secretion of the second hormone described by the concentration of the first hormone transformed by a time-lagged polynomial transfer function. Each hormone concentration time series was altered by adding random noise with SD equal to 1% of the mean concentration for the series.
[b] Denotes the coupling of the concentration of hormone *A* to the secretion of hormone *B*.
[c] Corresponds to the maximal strength of coupling.

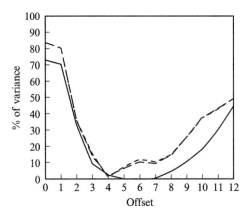

FIG. 4. The percentage of the variance for the secretion of hormone B accounted for by the concentration of hormone A modified by the polynomial transfer function, as a function of the degree of offset (see Methods). The solid line represents a linear transfer function, the long dashed line a quadratic function, and the short dashed line a cubic function. The concentration of hormone A was able to define up to 84% of the variance of the secretion of hormone B, with an offset of 0 data points.

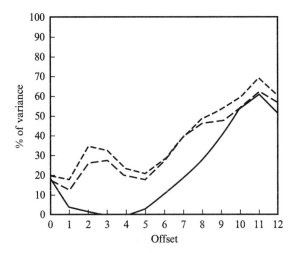

FIG. 5. The percentage of the variance for the secretion of hormone A accounted for by the concentration of hormone B modified by the polynomial transfer function, as a function of the degree of offset (see Methods). The symbols are defined as per Fig. 4.

Variability of the Strength of Coupling Caused by Deviation in Sample Collection Onset

To examine the reproducibility of the primary outcome, 10 different hormone concentration–time series pairs (hormones *A* and *B*) were created, each pair beginning at a distinct time point and modified by adding noise with SD equal to 1% of the mean. The coupling of *A* to *B* and vice versa was measured for each pair to create a distribution of values. The mean ±SD of the strength of coupling is shown in Table II.

Effect of Noise Increase on the Strength of Coupling

Increasing noise was added to each hormone time series. The effect of increasing noise on the measured strength of coupling is outlined in Table III and Fig. 6.

Effect of Half-Life Variability on the Predicted Strength of Coupling

The half-lives of hormones *A*, *B*, and *C* are known precisely, and are derived from the coupled differential equations describing this hormone network system. The hormone half-life is used in the deconvolution analysis to calculate the secretion rate of a given hormone, and usually is not known precisely, but rather is estimated from population physiology studies. To measure the effect of using a half-life in this calculation that did not precisely match the true hormonal half-life, the half-life specified for the deconvolution analysis was either increased or decreased by a fixed percentage. The effects on the percent of variance accounted for are shown in Fig. 7.

TABLE II
REPRODUCIBILITY OF STRENGTH OF COUPLING[a]

Order of transfer function	*A* to *B*[b]	*B* to *A*
Linear	73.3 ± 0.45	60.6 ± 1.3
Quadratic	84.2 ± 0.77	62.4 ± 1.3
Cubic	84.2 ± 0.78	69.2 ± 1.6

[a] The distribution of values for the percentage of the variance of the secretion of the second hormone described by the concentration of the first hormone transformed by a polynomial transfer function, using 10 paired *A–B* series modified by adding random noise with SD equal to 1% of the mean concentration for the series.

[b] Denotes the coupling of the concentration of hormone *A* to the secretion of hormone *B*.

TABLE III
EFFECT OF INCREASING MEASUREMENT ERROR (NOISE)[a]

Order of transfer function	Noise[b]	A to B^c	B to A	A to C	C to A	B to C	C to B
Linear	1	73	61	91	62	90	52
	5	70	59	84	60	85	50
	10	67	55	75	52	73	46
	15	63	47	62	49	63	42
	20	52	44	43	38	45	34
Quadratic	1	84	62	93	61	94	54
	5	81	61	85	59	88	51
	10	78	55	75	52	74	47
	15	75	48	62	49	64	44
	20	59	44	43	37	48	36
Cubic	1	84	69	95	67	94	54
	5	81	67	88	64	88	53
	10	78	61	76	59	76	50
	15	76	54	66	52	66	45
	20	58	48	46	44	48	40

[a] The percentage of the variance of the secretion of the second hormone described by the concentration of the first hormone transformed by a time-lagged polynomial transfer function.
[b] Random noise added to series, with noise SD equal to specified percent of the mean of the hormone concentration series.
[c] Denotes the coupling of the concentration of hormone A to the secretion of hormone B.

Reconstruction of Hormone Coupling

The procedure for determining suitable transfer functions described in Methods *de facto* reconstructs the original relationships between system components. This is exemplified in Fig. 8, which depicts plots of derived linear, quadratic, and cubic transferred functions for the coupling "A to Sec$B$" with random noise of 1% SD added to the model output data (results of the fitting procedure outcome are reported in Table I). The original control Hill function is shown for comparison. The difference in the fit between the linear and quadratic transfer functions is evident. Conversely, the cubic polynomial provides only moderate improvement over the quadratic fit. As expected, low-order polynomials can describe the S-shaped Hill function only over a limited part of its domain of definition (Methods). In this particular case, the quadratic and cubic curves almost coincide with the control over most of the apparent physiological range of A, but cannot account for the "A to Sec$B$" relationship for large nonphysiological concentrations of the stimulating hormone A.

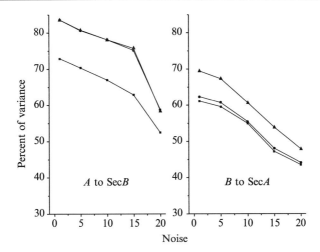

Fig. 6. Effect of signal noise on measured coupling. The percentage of the variance accounted for by the derived polynomial transfer function is shown for A to SecB (left) and B to SecA (right). Percent noise describes the addition of random noise with SD equal to a given percentage of the mean of the concentration of the hormone series, as a function of increasing signal noise. The order of the polynomial transfer function is ■, linear; ●, quadratic; ▲, cubic.

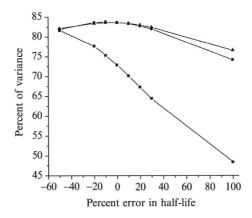

Fig. 7. Effect of error in assumed hormonal half-life on signal coupling. The percentage of the variance accounted for by the derived polynomial transfer function is shown for A to SecB, as a function of under- or overestimation of half-life. The symbols are as described for Fig. 6.

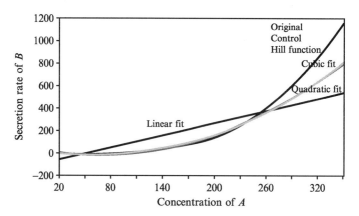

FIG. 8. Comparison between derived linear, quadratic, and cubic transfer functions for A to Sec$B$ and the original control Hill function (Fig. 1) over the apparent physiological concentration range of A.

Discussion

Endocrine systems typically involve multiple feedforward and feedback relationships that control the release of the involved hormones. A tool capable of measuring the strength of coupling between two hormones would have tremendous value for the investigation of normal hormonal physiology and for the investigation of various pathophysiological states. For instance, in the female reproductive system, one could hypothesize that the ability of luteinizing hormone (LH) to stimulate ovarian steroid production might be altered in aging or in polycystic ovarian syndrome. Statistical methods such as cross-correlation, cross-approximate entropy, and peak coincidence cannot precisely measure the strength of the coupling between the involved hormones on a minute-to-minute basis. The novel approach described previously has the advantage of being specifically designed to utilize dynamic features typical to endocrine feedback networks. Therefore, it is expected to be more discriminative.

The calculation of the strength of coupling between two hormones X and Y is based on the assumption that the secretion of Y is a dynamic variable, which depends on current (or previous) concentrations of X. The method starts with a known deconvolution procedure to determine the secretion rate of Y. Then, by a least-square fit it predicts the functional relationship (a transfer function in terms of low-order polynomial) between the already derived secretion of Y and the (delayed) concentration of X. The measure of the strength of the $X \rightarrow Y$ coupling is the percentage of the overall variance of the secretion of Y that could be accounted for by the concentration of X, converted by the derived transfer function.

We have shown that polynomial transfer functions can define a measure of coupling between two hormones. Several points are worth emphasizing when the mathematical construct and results are viewed together. First, the measurement of the percent of the variance accounted for by the polynomial transfer function was consistent. The coefficient of variation for the measurements ranged from 0.06% to 2%. Second, the measurement is affected by the number of hormones (or inputs) involved in the control of a given hormone. In this model, hormone B is controlled only by hormone A, and the concentration of hormone A was able to define 84% of the variance of the secretion of hormone B. Hormone A is controlled by both hormones B and C, and the concentration of hormone B was able to account for only 69% of the variance of the secretion of hormone A. However, both the $B \rightarrow A$ and $B \rightarrow C \rightarrow A$ pathways are evident in the percentage variance versus lag graphs. Third, measurement error reduces the ability of polynomial transfer functions to define accurately the coupling relationship. Increasing data noise caused a reduction in the measured coupling, particularly when the noise SD was greater than 15% of the mean hormone concentration. As the hormone half-life was used to estimate the hormone secretion using waveform-independent deconvolution analysis, overestimation of the half-life estimate also reduced the strength of coupling measured. Underestimation of the half-life did not have a significant impact on the measured strength of coupling. However, significant underestimation of the half-life ($\geq 20\%$) did induce a shift in the offset at which maximal coupling was observed. For example, for A to Sec$B$, the percent of the variance accounted for was greater at offset 1 compared to offset 0 if 50% underestimation of the half-life of B is assumed (data not shown). On the other hand, the actual half-life yield maximum coupling at offset 0, as would be predicted from the model (since A drives the secretion of B directly). The change in the maximum coupling offset is expected as the secretion profile generated by deconvolution analysis is shifted rightward when the half-life is underestimated.

There are several limitations worth mentioning. First, coupling does not imply a direct causal relationship. These data were derived from a known (synthetic) hormonal system where the coupling is defined mathematically, and it is known how the hormone concentrations affect the other components of the system. Thus, we can say precisely that the concentration of hormone A accounts for a given percentage of the secretion of hormone B. If the biological model were unknown, finding that hormone X is coupled to hormone Y would not imply causation (i.e., it does not follow that hormone X controls the release of hormone Y). It remains plausible that a third unmeasured substance controls simultaneously the release of X and Y. Second, the hormones' half-lives were known precisely in this model. Overestimation of these parameters leads to a decay in the estimated strength of coupling.

The formalism presented here is for the coupling of two hormones. However, it can easily be expanded to include multiple input hormones concentration–time series, each with the possibility of a unique time lag, controlling one, or more, output hormones.

In summary, we have shown that polynomial transfer functions can consistently describe the strength of the relationship between two hormones, and that the ability to measure this coupling decreases with increasing data noise and with increasing error in the estimated hormone half-lives. We are hopeful that this technique will be useful for the investigation of normal hormonal physiology and pathophysiology.

Acknowledgments

The authors acknowledge the support of the National Science Foundation Science and Technology Center for Biological Timing at the University of Virginia (NSF DIR-8920162) (M.L.J., W.S.E.), the General Clinical Research Center at the University of Virginia (NIH RR-00847) (M.L.J., W.S.E.), the University of Maryland at Baltimore Center for Fluorescence Spectroscopy (NIH RR-08119) (M.L.J.), and grant support by NIH K25 HD014744 (LSF) and F32 AG21365-01 (C.R.F.).

[7] Numerical Estimation of HbA$_{1c}$ from Routine Self-Monitoring Data in People with Type 1 and Type 2 Diabetes Mellitus

By BORIS P. KOVATCHEV and DANIEL J. COX

Introduction

Diabetes is a complex of disorders, characterized by a common final element of hyperglycemia, that arise from, and are determined in their progress by mechanisms acting at all levels of biosystem organization— from molecular to human behavior. Diabetes mellitus has two major types: Type 1 (T1DM) caused by autoimmune destruction of insulin-producing pancreatic beta cells, and Type 2 (T2DM), caused by defective insulin action (insulin resistance) combined with progressive loss of insulin secretion. Sixteen million people are currently afflicted by diabetes in the United States with epidemic increases now occurring. From 1990 to 1998 a one-third increase in diabetes in U.S. adults occurred. The risks and costs of diabetes (over $100 billion/year) come from its chronic complications in four major areas: retinal disease, which is the leading cause of adult blindness; renal disease, which represents half of all kidney failures; neuropathy, which predisposes to over 65,000 amputations each year; and cardiovascular disease, which in diabetics is two to four times more common than in those without diabetes. Cardiovascular disease in diabetes is also more morbid and more lethal, and benefits less by modern interventions such as bypass surgery or stents. Intensive treatment with insulin and with oral medication to nearly normal levels of glycemia markedly reduces the chronic complications of diabetes (except cardiovascular disease) in both types, but may risk severe and potentially life-threatening overtreatment due to hypoglycemia. State-of-the-art but imperfect replacement of insulin may reduce warning symptoms and hormonal defenses against hypoglycemia resulting in stupor, cognitive dysfunction, coma, driving accidents, seizures, brain damage, and sudden death.

Extensive studies, including the DCCT,[1] the Stockholm Diabetes Intervention Study,[2] and the UK Prospective Diabetes Study,[3] have repeatedly demonstrated that the most effective way to prevent long-term complications of diabetes is by strictly maintaining blood glucose (BG) levels within a normal range using intensive therapy. However, the same studies have also documented adverse effects of intensive therapy, the most acute of which is the increased risk of severe hypoglycemia (SH), a condition defined as severe neuroglycopenia that precludes self-treatment.[4–6] Since SH can result in brain abnormalities,[7] cognitive dysfunction,[8–10] accidents, coma, and even death,[5,11] hypoglycemia has been identified as the major barrier to improved glycemic control.[12–14] In short, a number of important

[1] The DCCT Research Group, *N. Engl. J. Med.* **329,** 978 (1993).

[2] P. Reichard and M. Phil, *Diabetes* **43,** 313 (1994).

[3] UK Prospective Diabetes Study Group (UKPDS), *Lancet* **352,** 837 (1998).

[4] R. Lorenz, C. Siebert, P. Cleary, J. Santiago, and S. Heyse, for the DCCT Research Group, *Diabetes* **37,** 3A (1988).

[5] The DCCT Research Group, *Am. J. Med.* **90,** 450 (1991).

[6] The Diabetes Control and Complications Trial Research Group, *Diabetes* **46,** 271 (1997).

[7] P. Perros, I. J. Deary, R. J. Sellar, J. J. Best, and B. M. Frier, *Diabetes Care* **20,** 1013 (1997).

[8] I. J. Deary, J. R. Crawford, D. A. Hepburn, S. J. Langan, L. M. Blackmore, and B. M. Frier, *Diabetes* **42,** 341 (1993).

[9] A. E. Gold, I. J. Deary, and B. M. Frier, *Diabetic Med.* **10,** 503 (1993).

[10] N. B. Lincoln, R. M. Faleiro, C. Kelly, B. A. Kirk, and W. J. Jeffcoate, *Diabetes Care* **19,** 656 (1996).

[11] Z. T. Bloomgarden, *Diabetes Care* **21,** 658 (1998).

[12] P. E. Cryer, J. N. Fisher, and H. Shamoon, *Diabetes Care* **17,** 734 (1994).

[13] P. E. Cryer, *Diabetes* **42,** 1691 (1993).

[14] P. E. Cryer, *Diabetes Metab. Res. Rev.* **15,** 42 (1999).

aspects of the pathogenesis of diabetes and its complications relate to *optimal control* of the insulin—carbohydrate balance and interactions. Thus, people with diabetes face the life-long *optimization problem* of maintaining strict glycemic control without increasing their risk of hypoglycemia. This optimization has to be based on data collection, data processing, and meaningful feedback available daily to individuals with T1DM and T2DM in their natural environment.

The classic marker of overall glycemic control in diabetes is glycosylated hemoglobin introduced more than 20 years ago,[15] and specifically its component HbA_{1c}, confirmed as the gold standard assay for both people with T1DM and T2DM.[16] Although HbA_{1c} determination is available in physician's offices and contemporary devices, such as A1c Now by Metrika,[17] offer at-home testing of HbA_{1c}, the tests require special efforts and are expensive. On the other hand, the standard at-home means of data collection include self-monitoring of blood glucose (SMBG) two or more times a day. Contemporary SMBG devices store hundreds of SMBG readings, connect to and upload data to a PC, and a number of software applications exist for data presentation and standard calculations, such as mean and standard deviation of BG.

Given the rapid development of the monitoring technology, surprisingly little attention is devoted to the processing and the interpretation of these data. Even the methods suggested for analysis of such rich data sets as continuous glucose monitoring (CGM) data are surprisingly "low-tech" compared to the technology itself, and are limited to summary tables of average BG, "modal day" profiles, and profiles with event markers.[18] A recent review of the existing discrete-time and continuous-time models of BG fluctuation and algorithmic strategies for BG control supports this observation.[19] This discrepancy is partially due to the fact that until recently, little utility (and modest success) has been seen in attempts to predict BG on the basis of previous readings.[20]

In particular, while in laboratory studies there is almost a perfect deterministic relationship between average blood glucose and HbA_{1c},[15] the relationship between average self-monitored BG (ASMBG) and HbA_{1c} is fairly uncertain and contaminated by randomness and unpredictability

[15] P. Aaby Svendsen, T. Lauritzen, U. Soegard, and J. Nerup, *Diabetologia* **23,** 403 (1982).
[16] J. V. Santiago, *Diabetes* **42,** 1549 (1993).
[17] Clinical Accuracy of A1C Now. Metrika: http://www.metrika.com/downloads/A1cNowProduct-PerformanceC.pdf.
[18] B. W. Bode, *Diabetes Technol. Ther.* **2**(Suppl. 1), S-35 (2000).
[19] D. R. L. Worthington, *Med. Inform.* **22,** 5 (1997).
[20] T. Bremer and D. A. Gough, *Diabetes* **48,** 445 (1999).

caused by the timing and frequency of SMBG.[21] For example, if a person measures his or her blood sugar predominantly after meals, ASMBG will be an overestimate of true average BG. In general, the numerical value of ASMBG is substantially influenced by readings taken during transient hypoglycemic or hyperglycemic episodes. However, such transient episodes, due to their short duration, do not account for substantial changes in HbA$_{1c}$. In addition, the BG measurement scale is asymmetric, which translates numerically into an excessive weight of high BG readings.[22] To compensate for that asymmetry we have developed mathematical models and analytical procedures that take into account specific properties of the BG scale.[23] On that basis, we created algorithms that (1) predicted 40–46% of SH episodes in the subsequent 6 months,[24–26] (2) identified 24-h periods of increased risk for hypoglycemia,[27] and (3) provided conversion tables containing 95% confidence intervals for translation of SMBG profiles into HbA$_{1c}$ and vice versa.[21] Most recently, we used these techniques to compare SMBG profiles in a large cohort of 600 patients with T1DM and T2DM, all of whom were taking insulin.[28]

We now apply these techniques to develop mathematical models estimating HbA$_{1c}$ from routine SMBG data. These models include corrections for underestimation due to frequent lower BGs as well as corrections for overestimation due to frequent higher BGs. We present two distinct models: Model 1, based only on SMBG and excluding any knowledge of a prior reference HbA$_{1c}$, and Model 2, combining a prior reference HbA$_{1c}$ with recent/current SMBG data. To ensure that these models can be generalized to the population level, we first develop and optimize the models using two "training" data sets including 181 subjects with T1DM and then test them in a large unrelated data set containing data for both T1DM and T2DM subjects.

[21] B. P. Kovatchev, D. J. Cox, M. Straume, and L. S. Farhy, *Methods Enzymol.* **321,** 410 (2000).

[22] B. P. Kovatchev, D. J. Cox, L. A. Gonder-Frederick, and W. L. Clarke, *Diabetes* **46**(Suppl. 1), A268 (1997).

[23] B. P. Kovatchev, D. J. Cox, L. A. Gonder-Frederick, and W. L. Clarke, *Diabetes Care* **20,** 1655 (1997).

[24] B. P. Kovatchev, D. J. Cox, L. A. Gonder-Frederick, and W. L. Clarke, *Diabetes* **47**(Suppl. 1), A107 (1998).

[25] B. P. Kovatchev, D. J. Cox, L. A. Gonder-Frederick, D. Young-Hyman, D. Schlundt, and W. L. Clarke, *Diabetes Care* **21,** 1870 (1998).

[26] B. P. Kovatchev, M. Straume, D. J. Cox, and L. S. Farhy, *J. Theor. Med.* **1,** 1 (2001).

[27] B. P. Kovatchev, D. J. Cox, L. S. Farhy, M. Straume, L. A. Gonder-Frederick, and W. L. Clarke, *J. Clin. Endocrinol. Metab.* **85,** 4287 (2000).

[28] B. P. Kovatchev, D. J. Cox, L. A. Gonder-Frederick, and W. L. Clarke, *Diabetes Technol. Ther.* **1**(3), 295 (2002).

Methods

Subjects

Training Data Set 1. Ninety-six patients with T1DM were diagnosed at least 2 years prior to the study. Forty-three of these patients reported at least two episodes of severe hypoglycemia in the past year and 53 patients reported no episodes during the same period. There were 38 males and 58 females. The mean age was 35 ± 8 years, mean duration of disease 16 ± 10 years, and mean insulin units/kg/day 0.58 ± 0.19.

Training Data Set 2. Eighty-five patients with T1DM were diagnosed at least 2 years prior to the study, all of whom reported SH episodes in the past year. There were 44 males and 41 females. The mean age was 44 ± 10 years, mean duration of disease 26 ± 11 years, and mean insulin units/kg/day 0.6 ± 0.2.

Test Data Set. The test data set that we used contains data for $N = 573$ subjects, 254 with T1DM and 319 with T2DM, all of whom used insulin to manage their diabetes. The two patient groups were matched by HbA_{1c}, number of SMBG readings/day, duration of diabetes, and gender. This is the same data set that was previously used to evaluate the accuracy of the relationship between HbA_{1c} and SMBG.[21] Table I presents demographic characteristics and comparison of T1DM versus T2DM subjects.

Data Collection Procedure

In all data sets SMBG was performed using Lifescan OneTouch II or OneTouch Profile meters.

Training Data Set 1. These subjects collected approximately 13,000 SMBG readings over a 40 to 45-day period. The frequency of SMBG was

TABLE I
DEMOGRAPHIC CHARACTERISTICS OF THE SUBJECTS IN THE TEST DATA SET

Variable	T1DM [mean (SD)]	T2DM [mean (SD)]	p level
Age (years)	38.0 (13.4)	58.1 (9.4)	<0.001
Gender: male/female	136/141	157/166	Ns
Baseline HbA_{1c}	9.74 (1.3)	9.85 (1.3)	Ns
HbA_{1c} at month 6	8.77 (1.1)	8.98 (1.3)	0.04
Duration of diabetes (years)	14.6 (9.8)	13.5 (7.6)	Ns
Age at onset (years)	23.4 (12.8)	44.6 (10.4)	<0.001
Number of SMBG readings/subject/day	3.2 (1.1)	2.9 (0.9)	<0.005

approximately three reading/day. During this period of time a single HbA_{1c} assay was performed. The mean HbA_{1c} was $8.6 \pm 1.8\%$. This data set was used as a training data set for Model 1 utilizing SMBG data only, without knowledge of a prior HbA_{1c}.

Training Data Set 2. These subjects collected approximately 75,500 SMBG readings over 6 months. The frequency of SMBG was four to five readings per day. Two HbA_{1c} assays were performed—one in the beginning and one at the end of this 6-month period. The mean baseline HbA_{1c} was $7.7 \pm 1.1\%$, and the mean 6-month HbA_{1c} was $7.4 \pm 1\%$ (6-month HbA_{1c} was available for 60 subjects). This data set was used as a training data set for Model 2 using prior HbA_{1c} and recent/current SMBG data.

Test Data Set. These data were collected by Amylin Pharmaceuticals, San Diego, CA, and included 6–8 months of SMBG data (approximately 300,000 readings), accompanied by baseline and second HbA_{1c} taken approximately 6 months later. The exact dates of HbA_{1c} determinations were not known. The subjects were participating in a clinical trial investigating the effects of pramlintide (in doses of 60–120 μg) on metabolic control. The subjects' use of pramlintide was randomized across the T1DM and T2DM groups. For the first 6 months of the study the average HbA_{1c} declined significantly in both T1DM and T2DM groups, perhaps due to the use of medication, which is beyond the scope of this presentation (Table I). This relatively rapid change in HbA_{1c} allowed for a better estimation of the predictive ability of our models.

Estimation Procedure

In addition to the average of SMBG readings, computed in millimoles/liter, we used two previously introduced measures—the low and high BG indices (LBGI and HBGI[26]) derived from self-monitoring data. Several different functions were considered for description of the relationship between SMBG data and HbA_{1c}. Optimal, in terms of accuracy and simplicity of computation, appeared to be a linear function of the average of SMBG readings, LBGI and HBGI, if no prior HbA_{1c} reading was used, and another linear function of a prior HbA_{1c} and the HBGI. Nonlinear relationships did not enhance the goodness of fit of the models and therefore are not considered in detail here.

The computation of the LBGI and HBGI goes through the following steps: (1) symmetrization of the BG measurement scale, (2) assignment of a risk value to each SMBG reading, and (3) computing LBGI and HBGI. In brief these steps are performed as follows:

1. *Symmetrization of the BG scale:* A nonlinear transformation is applied to the BG measurements scale to map the entire BG range

(1.1–33.3 mmol/liter or 20–600 mg/dl) to a symmetric interval $(-\sqrt{10},$ $\sqrt{10})$. The point 6.25 mmol/liter (112.5 mg/dl), which is the clinical center[23] of the BG scale is mapped to 0. As previously reported,[23,26] the analytical form of this transformation is as follows:

$$f(BG, \alpha, \beta) = \{[\ln (BG)]^{\alpha} - \beta\}, \qquad \alpha, \ \beta > 0$$

where the parameters are estimated as $\alpha = 1.026$, $\beta = 1.861$, and $\gamma = 1.794$ if BG is measured in millimoles/liter and as $\alpha = 1.084$, $\beta = 5.381$, and $\gamma = 1.509$ if BG is measured in milligrams/deciliter.

2. *Assignment of a risk value to each SMBG reading:* After fixing the parameters of $f(BG)$ depending on the measurement scale that is being used, we define the quadratic risk function $r(BG) = 10f(BG)^2$. The function $r(BG)$ ranges from 0 to 100. Its minimum value is achieved at $BG = 6.25$ mmol/liter (112.5 mg/dl), a safe euglycemic BG reading, while its maximum is reached at the extreme ends of the BG scale. Thus, $r(BG)$ can be interpreted as a measure of the risk associated with a certain BG level. The left branch of this parabola identifies the risk of hypoglycemia, while the right branch identifies the risk of hyperglycemia.

3. *Computing LBGI and HBGI:* Let $x_1, x_2, \ldots, x_n$ be a series of n BG readings, and let

$$rl(BG) = r(BG) \quad \text{if} \ \ f(BG) < 0 \ \text{otherwise}$$

$$rh(BG) = r(BG) \quad \text{if} \ \ f(BG) > 0 \ \text{and} \ 0 \ \text{otherwise}$$

The low blood glucose [risk] index (LBGI) and the high BG [risk] index (HBGI) are then defined as

$$LBGI = \frac{1}{n}\sum_{i=1}^{n} rl(x_i) \quad \text{and} \quad HBGI = \frac{1}{n}\sum_{i=1}^{n} rh(x_i)$$

In other words, the LBGI is a nonnegative quantity that increases when the number and/or extent of low BG readings increase. Similarly, the HBGI increases when the number and/or extent of high BG readings increase. Each index and their sum LBGI + HBGI have a theoretical upper limit of 100. It has been demonstrated that the LBGI is a predictor of severe hypoglycemia,[24,25] while the HBGI has been associated with HbA_{1c}.[21] The mathematics behind the LBGI and HBGI has been previously presented in detail.[26] In the context of this presentation we will use the LBGI to increase the weight of low BG readings in the numerical estimation of HbA_{1c}, and we will use the high BG index to decrease the weight of high transient BGs into the estimation of HbA_{1c}.

Results

Training Data Set 1: Model 1: No Prior HbA$_{1c}$

ASMBG, LBGI, and HBGI were computed from SMBG readings taken over the 45 days preceding HbA$_{1c}$ determination. A linear regression model with a dependent variable HbA$_{1c}$, predictors ASMBG (coefficient = 0.81), LBGI (coefficient = 0.06), and HBGI (coefficient = -0.15), and intercept of 1.87 was highly significant ($F = 31.2, p < 0.0001$), < 0.0001), and resulted in $R^2 = 0.51$ and multiple $R = 0.71$. Analysis of the residuals of this model showed a close to normal distribution of the residuals with a mean of 0 and SD = 1.2. Therefore we accepted that this model described the data well.

Training Data Set 2: Model 2 Using a Prior HbA$_{1c}$

Baseline HbA$_{1c}$ was measured 6 months prior to current HbA$_{1c}$ and HBGI was computed from SMBG readings taken over the 45 days preceding current HbA$_{1c}$. A linear model with a dependent variable current HbA$_{1c}$, predictors baseline HbA$_{1c}$ (coefficient = 0.68), HBGI (coefficient = 0.05), and intercept of 1.55 was highly significant ($F = 41.7, p < 0.0001$), and resulted in $R^2 = 0.76$ and multiple $R = 0.87$. Analysis of the residuals of this model showed a close to normal distribution with a mean of 0 and SD = 0.5. Therefore we can accept that this model described the data well.

Test Data Set: Accuracy of Models 1 and 2

For each subject, a 45-day subset of his or her SMBG reading was selected. Since in this data set the exact time of HbA$_{1c}$ assays was not known, the SMBG time periods for each subject could be selected to precede the second HbA$_{1c}$ determination only approximately. The two models derived in the two training data sets were applied, without changing their coefficients, to the test data set.

Several standard criteria were used to assess the accuracy of Models 1 and 2:

1. Absolute deviation (AERR) of estimated from measured HbA$_{1c}$.
2. Absolute percent deviation (PERR) of estimated from measured HbA$_{1c}$.
3. Coefficient of variation CV = SD of errors/HbA$_{1c}$.
4. Percent estimates within 20% of measured HbA$_{1c}$ (HIT 20).
5. Percent readings within 10 of measured HbA$_{1c}$ (HIT 10).
6. Percent readings outside of a 25% zone around measured HbA$_{1c}$ (MISS 25).

It is important to note that it is not appropriate to evaluate the accuracy of the models in the test data set using regression-type criteria, such as R^2 or F/p values from an ANOVA table. This is because the parameter estimates were derived from other unrelated data sets (the training data) and are only applied to this test data set. Thus, statistical assumptions for the underlying model are violated (for example, in the test data set the sum of the residuals will not be zero) and therefore R^2, F, and p lose their statistical meaning.

Table II presents results from the evaluation of Models 1 and 2 with data from the test data set for subjects with T1DM and T2DM, respectively. The table also includes three additional columns. The third column presents the accuracy of the estimation if ASMBG (in millimoles/liter) was taken as an estimate of HbA_{1c}. The fourth column presents the accuracy of evaluation of current HbA_{1c} if the baseline determination of HbA_{1c} was taken as an estimate. The last column of the table presents the p value of a 1×4 ANOVA comparing the accuracy of Model 1, Model 2, ASMBG, and baseline HbA_{1c}.

Table II demonstrate that for both T1DM and T2DM Model 2 is a little better overall estimate of HbA_{1c} compared to Model 1. However, both Models 1 and 2 are substantially better estimates of HbA_{1c} than its baseline value or than the average BG. This is especially true for the percentage estimates that fell outside of the 25% accuracy zone.

TABLE II
ACCURACY IN T1DM AND T2DM

	Model 1	Model 2	ASMBG	Baseline HbA_{1c}	p value
T1DM ($N = 254$)					
AERR	0.77	0.61	1.68	1.1	<0.001
PERR (%)	8.3	7.1	19.4	12.8	<0.001
CV (%)	10.4	8.9	17	10.9	<0.001
HIT20 (%)	96.5	95.7	61.0	81.0	<0.001
HIT10 (%)	65.4	75.5	29.9	48.2	<0.001
MISS 25 (%)	2.4	1.6	28.4	9.9	<0.001
T2DM ($N = 319$)					
AERR	0.72	0.57	1.92	0.87	<0.001
PERR (%)	7.6	6.4	20.9	11.7	<0.001
CV (%)	10	8	18	9.8	<0.001
HIT20 (%)	95.9	98.4	56.4	82.8	<0.001
HIT10 (%)	70.2	79.3	29.5	53.3	<0.001
MISS 25 (%)	1.2	0.6	36.7	8.2	<0.001

Test Data Set: Capturing Trends in HbA$_{1c}$

The ability of Models 1 and 2 to capture trends in HbA$_{1c}$ was evaluated by reviewing the T1DM and T2DM subjects who had a substantial change in their SMBG reading from the baseline to 6-month follow-up. From baseline to second HbA$_{1c}$ 34 T1DM and 34 T2DM subjects had absolute changes in their HbA$_{1c}$ equal to or greater than 2 units. Model 1 predicted 100% of such changes in both T1DM and T2DM. The power of Model 2 to predict such changes was diminished due to the inclusion of baseline HbA$_{1c}$ in the equation (which partially pulls the estimates back to the baseline value of HbA$_{1c}$) and was 71% in T1DM and 85% in T2DM.

Conclusions

We have presented models using SMBG to estimate HbA$_{1c}$, the most important and widely accepted marker of glycemic control of people with Type 1 or Type 2 diabetes. In parallel with the optimization of the estimating functions we studied the optimal duration of the SMBG data collection period and the optimal frequency of self-monitoring during that period. We concluded that (1) the optimal SMBG data collection period is 45 days, and (2) the optimal frequency of SMBG is three reading per day. Two optimal HbA$_{1c}$ linear functions were developed: *Model 1*—using only SMBG data, and *Model 2*—using SMBG data plus an HbA$_{1c}$ reading taken approximately 6 months prior to the HbA$_{1c}$ that is being predicted.

Before arriving at these two models, we tested a number of alternative approaches, such as selecting specific times of the day (postprandial reading) for evaluation of HbA$_{1c}$, different weighting of SMBG readings according to the elapsed time between each SMBG reading and HbA$_{1c}$ determination, separate evaluation of subjects with a different average blood glucose to HbA$_{1c}$ ratio, etc. While some of these alternative approaches achieved better results than the two linear functions proposed previously, none was better overall. Experiments with different weighting of SMBG reading dependent on the elapsed time between SMBG and HbA$_{1c}$ assay (such as weighting higher more proximal results) did not yield a better prediction of HbA$_{1c}$. Specifically, we tested (1) weighting higher the more proximal SMBG, (2) weighting higher more prolonged high BG events, and (3) calibrating the high BG index with an earlier HbA$_{1c}$. Finally, incorporating demographic variables, such as age, duration of diabetes, and gender, did not improve additionally the prediction of HbA$_{1c}$.

We used our data to verify literature reports claiming that HbA$_{1c}$ was more closely related to preprandial than postprandial plasma/blood glucose levels.[29] Since there were also reports claiming the opposite,[30] this result remained unclear.[31] In our data we found that HbA_{1c} was most associated (in terms of correlation) with SMBG readings taken in the afternoon hours—1–6 PM and least associated with SMBG readings taken at night (2–7 AM). Specifically, the correlations of afternoon (presumably postprandial) readings with HbA_{1c} were 0.50 for T1DM and 0.74 for T2DM, while the correlations of nightly (presumably preprandial) readings with HbA_{1c} were 0.32 and 0.65 for T1DM and T2DM, respectively (all correlations were significant at $p = 0.005$). However, taking only postprandial SMBG readings did not improve the prediction of HbA_{1c}. On the contrary, the prediction became worse if the contribution of *any* time blocks throughout the day was ignored. Different weighting of different hours throughout the day did not improve significantly the prediction of HbA_{1c}, thus this additional complication of the models was not justified.

The direct association between HbA_{1c} and ASMBG was substantially stronger in T2DM compared to T1DM, even though the two subject groups had similar HbA_{1c} (Table I). In terms of direct correlation, in T1DM the coefficient was 0.6 while in T2DM the coefficient was 0.75 in the test data set, which was reflected by a (nonsignificantly) better prediction of HbA_{1c} among T2DM subjects. This difference between T1DM and T2DM was perhaps due to the greater variability of BG fluctuations in T1DM. As we demonstrated in previously reported analyses of this same data set, patients with T1DM had (1) a greater risk for SH as quantified by the low BG index, (2) faster descent into hypoglycemia, and (3) greater BG irregularity, all p levels <0.0001.[28]

It is essential to note that in addition to estimating HbA_{1c}, the broader goal of the models presented here is to evaluate the status of patients' glycemic control. Although HbA_{1c} is the accepted "gold standard" for evaluation of glycemic control, currently it is unclear whether another measure, such as ASMBG or high BG index, would not be a similar to, or a better predictor of long-term complications in diabetes than HbA_{1c}. Until this issue is clarified, the goal of our models will be to estimate HbA_{1c}. To approximate as closely as possible future real applications of these models we proceeded as follows. (1) First, several optimal functions using different independent variables, optimal duration, and optimal frequency of SMBG were derived from two *training data sets,* collected in our previous studies, involving patients with T1DM. (2) Then all coefficients were fixed and the

[29] E. Bonora, F. Calcaterra, S. Lombardi, N. Bonfante, G. Formentini, R. C. Bonadonna, and M. Muggeo, *Diabetes Care* **24**, 2023 (2001).

[30] A. Avignon, A. Radauceanu, and L. Monnier, *Diabetes Care* **20**, 1822 (1997).

[31] S. Caputo, D. Pitocco, V. Ruotolo, and G. Ghirlanda, *Diabetes Care* **24**, 2011 (2001).

algorithm was applied to the much larger *test data set* containing data for both T1DM and T2DM subjects collected under very different conditions in a clinical trial conducted by Amylin Pharmaceuticals. (3) Finally, detailed estimation of the preciseness of the algorithm for various optimal functions was made using only the test data.

This separation of *training* and *test* data sets allows us to assume that the estimated preciseness of the algorithm can be generalized to any other data of subjects with T1DM or T2DM. Moreover, since the Amylin data (test data set) were collected from subjects who were undergoing treatment to lower their HbA$_{1c}$, and therefore exhibited an unusually large variation of their HbA$_{1c}$ over the 6-month period of observation, we could assume that Model 1, which did not depend on a prior HbA$_{1c}$ reading, would be especially predictive of large and unusually rapid changes in HbA$_{1c}$. This assumption was confirmed by the observation that for subjects whose HbA$_{1c}$ changed two or more units from their baseline reading ($N = 68$), the accuracy of Model 1 in predicting this change was 100% in both T1DM and T2DM, while the accuracy of Model 2 was only 71% and 85% in T1DM and T2DM, respectively. Thus, Model 1 would be most useful for patients whose goal is to optimize their HbA$_{1c}$.

The evaluation of the accuracy of HbA$_{1c}$ prediction in the test data set ($N = 573$ subjects) was done by several widely accepted criteria. The coefficients of variation of Model 1 were 10% for both T1DM and T2DM subjects and the coefficients of variation for Model 2 were below 9% for both subject groups. Since the College of American Pathology (CAP) reports percentage CVs of about 3–13% for A1c from proficiency testing by laboratories around the United States, we can conclude that the accuracy of both models was comparable to the accuracy of direct measurement of HbA$_{1c}$ across different laboratories. In addition, both Models 1 and 2 provided a better estimation of current HbA$_{1c}$ than HbA$_{1c}$ reading taken 6 months before, and a substantially better estimation than the ASMBG. Since prior HbA$_{1c}$ readings and ASMBG are widely used in the clinical practice as measures of patients' glycemic control, a suggestion will be to process first these measures through Model 1 or Model 2, and then to use the result as an estimate of current HbA$_{1c}$. Comparing the models without and with a prior HbA$_{1c}$, we see that if a prior HbA$_{1c}$ is available for inclusion in the computations, the resulting Model 2 is generally better both in terms of R^2 and in terms of residual error. However, when subjects' HbA$_{1c}$ changed rapidly, Model 1 was superior in tracking these changes.

We conclude that SMBG data contain valuable information about individuals' glycemic control that, using uncomplicated computation, can be accurately translated into HbA$_{1c}$—the most accepted marker of glycemic control. Since SMBG data are recorded and stored by most contemporary home-monitoring devices, incorporating an appropriate computation and presentation of results would take the utility of such devices well beyond the currently available display of a momentary snapshot of patients' glucose status.

Acknowledgments

This study was supported by a grant and material support from LifeScan Inc, Milpitas, CA, and by Grant DK 51562 from the National Institutes of Health. We are thankful to Amylin Pharmaceuticals, San Diego, CA, for sharing their database with us.

[8] Wavelet Modeling and Processing of Nasal Airflow Traces

By Mario Peruggia, Junfeng Sun, Michael Mullins, and Paul Suratt

Introduction

The underlying motivation for the work presented in this chapter is the desire to develop sound methodological instruments for the analysis of breathing during sleep. In subjects undergoing a polysomnography study, signals reflecting breathing and sleep are monitored and recorded at regular time intervals throughout an entire night. Typically, tracings from sleep studies are reviewed by laboratory technicians who are trained to recognize abnormalities in patterns of the signals. To go beyond this approach, which is essentially of an empirical nature and subject to the vagaries of subjective interpretation, one must be able to extract precise quantitative measures from the recorded signals. To fulfill this goal, we have completed the initial developmental phases of a computer algorithm for the processing of nasal airflow data. The fully automated algorithm is novel in several respects and its output can help to overcome the limitation of a categorical classification of breathing episodes.

Our analysis of nasal airflow began with a careful review of the tracings extracted from sleep studies conducted on a group of 20 patients who were referred to the Sleep Disorders Laboratory of the University of Virginia Medical Center for evaluation of possible sleep apnea. First, we made the fundamental observation that nasal airflow varies continuously, so that the usual classification of breathing abnormalities into a finite number of categories[1,2] is unnecessarily restrictive. Next, we proceeded to determine aspects of the signal that can be measured on a continuous scale.

Specifically, we noticed that the majority of sleep cycles are made up of an inspiration period (I), followed by an expiration period (E), which in turn contains a ventilatory pause of nearly zero flow (NF period, for no-flow). While the NF periods usually occur at the end of the E periods, we also noticed that they may occasionally occur at other points within the breathing cycle. Incidentally, the discovery that the NF periods are an integral part of the breathing cycles was triggered by the computational difficulties encountered in an initial attempt to develop an algorithm for the automated detection of breathing cycles that did not account explicitly for their presence.

The algorithm based on wavelet shrinkage estimation that we developed allows us to obtain precise measurements of the duration of the I/E/NF periods. On the basis of these durations derived measures of breathing that vary on a continuous scale and appear to be related to the characteristics and degree of severity of sleep-disordered breathing among subjects can be constructed. Some of these derived measures incorporate explicitly the durations of the NF periods. While the main goal of this chapter is to illustrate the mechanics and performance of the algorithm, we will also present a brief example of a derived measure of breathing and its possible application.

Wavelet Modeling of Airflow Traces

The development of an automated procedure to identify the end points of the I/E/NF periods is rather challenging, especially for the highly irregular airflow signals of diseased subjects. Our algorithm achieves this goal by denoising and smoothing the signal using wavelet shrinkage based on two different types of discrete wavelet transforms. The bases for the two transforms are chosen so as to obtain estimates that enhance different and complementary features of the signal. This is an important and novel methodological point: rather than pursuing a single, jack-of-all-trades estimate that does only a "passable" job of describing all important features of the signal, we seek two separate, specialized types of estimates each of which does a "good" job of describing complementary sets of important features and we exploit them in a combination that lets them borrow strength from one another.

---

[1] J. Hosselet, I. Ayappa, R. G. Norman, A. C. Krieger, and D. M. Rapoport, *Am. J. Respir. Crit. Care Med.* **163,** 398 (2001).

[2] W. W. Flemons, D. Buysse, S. Redline, A. Pack, K. Strohl, J. Wheatley, T. Young, N. Douglas, P. Levy, W. McNicholas, J. Fleetham, D. White, W. Schmidt-Nowarra, D. Carley, and J. Romaniuk, *Sleep* **22,** 667 (1999).

A Review of Wavelets

Before describing our algorithm, we review some basic results in wavelet analysis. There are very many references that present the principles of wavelet analysis, providing varying levels of mathematical detail. (A few references, roughly in order of increasing technical difficulty, are Bruce and Gao,[3] Vidakovic,[4] and Chui.[5]) In our exposition, we mainly follow the presentation, terminology, and notation of Bruce and Gao.[3] This choice is motivated by the fact that this approach[3] is not overly technical and that the book carefully describes how to perform wavelet analysis via the module `wavelets` of the statistical programming language S-Plus by Insightful Corporation (www.insightful.com). S-Plus is the programming environment that we used to implement our algorithm. A clear introduction to its use is given in Krause and Olson.[6]

Wavelets can be thought of as building blocks for approximating functions, in much the same way as sine and cosine waves are used to approximate functions in Fourier analysis. There are, however, fundamental differences in the properties enjoyed by the two resulting types of approximations. The functions $\sin(t)$ and $\cos(t)$ are periodic functions of period $P = 2\pi$, i.e., $\sin(t) = \sin(t + 2\pi)$ and $\cos(t) = \cos(t + 2\pi)$, for all $t \in (-\infty, \infty)$. (For simplicity of exposition the argument t is assumed to represent time.) Under suitable regularity conditions, an approximation of a periodic function $f(t)$ of period $P = 2T$ is given by

$$f(t) \approx \frac{a_0}{2} + \sum_{n=1}^{N} \left(a_n \cos\frac{n\pi t}{T} + b_n \sin\frac{n\pi t}{T} \right) \tag{1}$$

where the coefficients of the trigonometric polynomial on the right-hand side are defined by

$$a_n = \frac{1}{T} \int_{-T}^{T} f(t)\cos\frac{n\pi t}{T}\,dt, \quad n = 0, 1, \ldots$$

and

$$b_n = \frac{1}{T} \int_{-T}^{T} f(t)\sin\frac{n\pi t}{T}\,dt, \quad n = 1, 2, \ldots$$

[3] A. Bruce and H.-Y. Gao, "Applied Wavelet Analysis With S-Plus." Springer-Verlag, New York, 1996.

[4] B. Vidakovic, "Statistical Modeling by Wavelets." Wiley, New York, 1999.

[5] K. C. Chui, "An Introduction to Wavelets." Academic Press, San Diego, 1992.

[6] A. Krause and M. Olson, "The Basics of S-Plus," 3rd Ed. Springer-Verlag, New York, 2002.

The expression in Eq. (1) is a partial sum of an infinite expansion called the *Fourier series* of *f*.

The rescaled sine and cosine functions 1, $\cos(\pi t/T)$, $\cos(2\pi t/T)$, ... , $\sin(\pi t/T)$, $\sin(2\pi t/T)$, ... constitute an orthogonal basis on $[-T,T]$.[4] (The use of orthogonal bases to approximate functions greatly simplifies the calculation of the coefficients in the approximation and further theoretical developments.) One important aspect to notice is that the basis functions are all periodic and do not damp down to zero. As a consequence, approximation of a signal may often require many basis functions and rely on cancellation. This (undesirable) characteristic, shared by other classic approximations based on orthogonal bases (Hermite, Legendre, etc.), is referred to as *nonlocality*.[4]

Wavelets are different in this respect: they generate *local* bases that do a good job of describing signals that exhibit jumps, nonsmooth behavior, and features that change over time.[3] This is due to the fact that wavelet bases are generated by translating and rescaling primitive building blocks that are either identically zero outside a compact set (e.g., the *Haar* wavelets in the two top panels of Fig. 1 and the s8 *symmlet* wavelets in the two bottom panels of Fig. 1) or ones that damp down to zero. The wavelets in the figure were generated using the S-Plus function wavelet().[3]

As is the case with Fourier approximations, the signal is eventually represented by a linear combination of elements of an orthogonal basis. There are two pieces that play a fundamental role in the construction of a wavelet basis. A *father wavelet*, ϕ, satisfying the condition $\int \phi(t)dt = 1$ (e.g., panels 2 and 4 in Fig. 1) and a *mother wavelet*, ψ, satisfying the condition $\int \psi(t)dt = 0$ (e.g., panels 1 and 3 in Fig. 1).[3] A linear combination of rescaled and translated copies of a father wavelet is good at representing the smooth (low frequency) part of a signal. Linear combinations of rescaled and translated copies of a mother wavelet are good at representing the detail (high frequency) parts of a signal.

The separation of the smooth and detail components yields a *multiresolution approximation*[3] of a signal of the form

$$f(t) \approx S_J(t) + D_J(t) + D_{J-1}(t) + \ldots + D_1(t) \qquad (2)$$

where

$$S_j(t) = \sum_k s_{j,k} \phi_{j,k}(t)$$

and

$$D_j(t) = \sum_k d_{j,k} \psi_{j,k}(t)$$

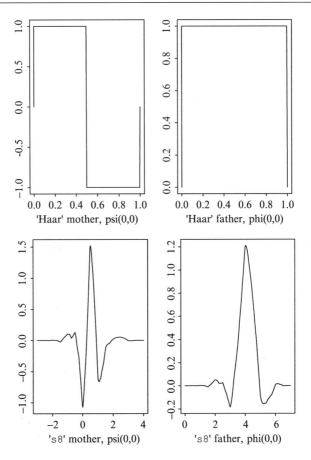

Fig. 1. The Haar and s8 mother and father wavelets.

The functions $\phi_{j,k}(t)$ and $\psi_{j,k}(t)$ form an orthogonal basis[3] and are computed from ϕ and ψ via translation and rescaling as

$$\phi_{j,k}(t) = 2^{-j/2}\phi(2^{-j}t - k)$$

and

$$\psi_{j,k}(t) = 2^{-j/2}\psi(2^{-j}t - k)$$

The coefficients of the expansion are called the *wavelet transform coefficients* of the signal. Their approximate values are given by

$$s_{j,k} \approx \int \phi_{j,k}(t)f(t)dt$$

and

$$d_{j,k} \approx \int \psi_{j,k}(t)f(t)dt$$

In practice, as is the case in the nasal airflow application, the signal is sampled at a finite number of equally spaced time points. Given a discrete signal of this sort, the *discrete wavelet transform* (DWT) establishes a correspondence between the sampled values and the wavelet transform coefficients.[3] Computationally, the coefficients can be readily derived using a *cascade* or *pyramid* calculation scheme.[3]

Depending on the nature of the signal to be approximated, the use of one or another wavelet family might be more or less appropriate. Consider, for example, the sample signal denoted by "Data" in the top row of the two panels of Fig. 2. The first half of the signal is essentially a step function while the second half exhibits a smooth sinusoidal behavior. Due to its "boxy" nature, the Haar wavelet family is an ideal candidate for approximating the first half of the signal but it does not perform so well with the second half. Vice versa, the s8 wavelet family is more appropriate for approximating the second rather than the first half of the signal. This can be

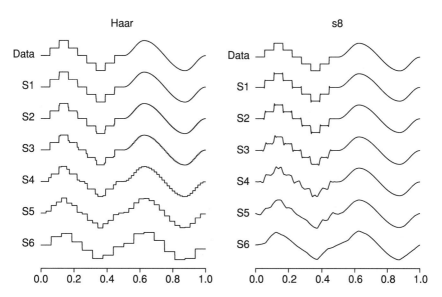

Fig. 2. Two approximations of a sample signal.

seen in Fig. 2, which displays two sets of multiresolution approximations of the sample signal produced with the S-Plus function $mra()$.[3] The approximations in the left panel are obtained using the Haar wavelet family and those in the right panel using the s8 wavelet family. There are several rows of approximations that become progressively more refined moving from the bottom to the top of the display. Using the same notation as in Eq. (2), they are obtained by progressively adding more and more detail signals to the smooth signal S_6. Precisely, $S_5 = S_6 + D_6$, $S_4 = S_6 + D_6 + D_5$, all the way to the sampled data that equals $S_6 + D_6 + D_5 + \ldots + D_1$. Visual inspection reveals the superiority of the Haar wavelet family approximation compared to the s8 wavelet family approximation over the first half of the signal (with the exception of the approximation at the jump points of the step function) and a reversal of performance over the second half. This observation provides the main motivation for the development of our algorithm.

In general, signals are measured with noise and one of the goals of a statistical analysis is to separate out the signal from the noise. Wavelet shrinkage estimation pursues this goal by shrinking the coefficients in the DWT toward zero and reconstructing the signal by inverting the DWT. In many cases this accomplishes the goal of reducing the noise while preserving the essential features of the signal. Various shrinkage methods are available with different theoretical properties.[3,4] A signal estimate computed according to this approach is usually referred to as a *wavelet shrinkage estimate*. The usefulness of techniques based on the use of the DWT to automate the analysis of medical signals is being recognized. For example, the use of the DWT to automate the detection of arousals during sleep has been documented.[7]

The S-Plus function $waveshrink()$ that we used to implement our algorithm has many optional arguments that control the way in which the shrinkage of the coefficients is performed.[3] Among these are the argument $wavelet$ that specifies what basis is used to calculate the DWT, the argument $shrink.fun$ that specifies the shrinkage function, and the argument $shrink.rule$ that specifies the shrinkage rule used to determine the shrinkage threshold used by the shrinkage function. The implementation of our algorithm makes use of wavelet shrinkage estimates of the airflow signal based on both the Haar and s8 bases. In all of the simulation studies presented we used the "hard" threshold function (a function that annihilates all coefficients at or below a certain level and leaves unaltered all coefficients above that level). In all cases we also set $shrink.rule$ equal to "minimax."[3]

[7] F. De Carli, L. Nobili, P. Gelcich, and F. Ferrillo, *Sleep* **22**, 561 (1999).

The performance of our algorithm is affected by the amount of smoothing that is performed. The argument `smooth.levels` in the function `waveshrink()` lets the user control the resolution levels to which shrinkage is applied. If `smooth.levels` is set equal to j, then only the detail coefficients at the j finest scales (i.e., the detail coefficients appearing in $D_1, \ldots, D_j$) are shrunk. The larger the value of j, the smoother the resulting estimate of the signal.

Typical Airflow Traces

The top panels in Fig. 3 display nasal airflow tracings recorded over intervals spanning 20 s for two of the 20 subjects under study. The signals were monitored every one-sixteenth of a second with nasal prongs attached to a pressure transducer (Validyne). An evaluation of the accuracy of this measurement technique is contained in Montserrat *et al.*[8] The left panel corresponds to a normal subject while the right panel corresponds to a subject diagnosed with sleep-disordered breathing. The dashed horizontal line denotes zero flow, with negative flow values corresponding to periods of inspiration, and positive flow values corresponding to periods of expiration.

The signal for the normal subject appears to be fairly regular and the I/E/NF periods are clearly delineated. (The graphic conventions used in the figure are explained in the last paragraph of the next section.) For example, the first complete cycle labeled 25 begins with an I period starting at 124.2188 s, followed by an E period starting at 126.3492 s, with an NF period starting at 128.4688 s and ending at 129.4688 s. Other cycles follow according to a rather predictable pattern. On the other hand, the signal for the diseased subject is much more unpredictable. For example, the stretch of tracing following the expiration starting at 24.4382 s is comprised of several unsuccessful attempts at a full inspiration coupled with suspected mouth breathing. A full recovery has not yet occurred by 40 s.

The irregular nature of the signal for Subject 17 is indicative of the types of obstacles we had to face and overcome to develop an automated procedure to identify the I/E/NF periods. On the basis of the traces for all 20 subjects we observed that, in general, the Haar wavelet basis provides a good representation of the signal in the zones of nearly zero flow (the NF periods) that occur mainly at the end of expiration and that the s8 wavelet basis provides a good representation of the signal in those zones where there is substantial flow (either negative, i.e., inspiration, or positive, i.e., expiration).

[8] J. M. Montserrat, R. Farre, E. Ballester, M. A. Felez, M. Pasto, and D. Navajas, *Am. J. Respir. Crit. Care Med.* **155,** 211 (1997).

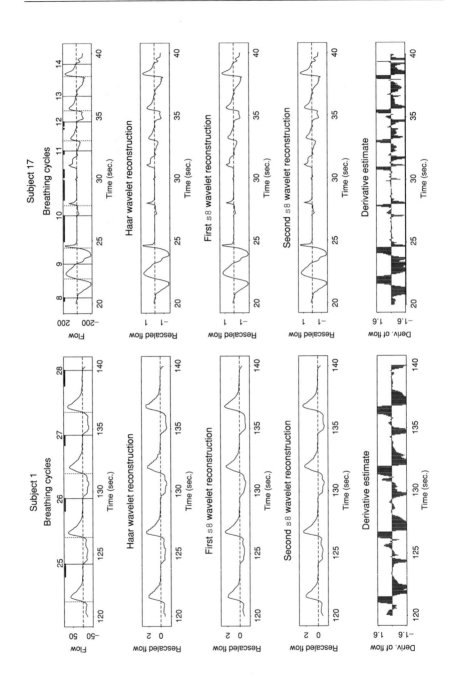

In addition, the s8 wavelet basis can be used to obtain estimates of the rate of change (derivative) of the signal. These estimates are more reliable in the zones of high flow where the s8 wavelet basis provides a good representation of the signal. However, as we will explain later, when used in combination with the Haar basis representation of the original signal, they prove to be useful even over the less reliable zones of little flow where they play a minor role in aiding the detection of the NF periods.

Implementation

On account of the previous observations, we developed an algorithm that employs wavelet shrinkage based on two different types of DWT (Haar and s8) to denoise and smooth the signal. This approach has the advantage that the strong points of either representation are used in a synergistic fashion. The algorithm can handle the great majority of patterns that occur in recorded airflow signals and identify the various components that make up the sleep cycles. Omitting a description of the rules that we introduced to deal with special patterns that occur rather rarely and other minor details, the basic framework of the algorithm can be simply summarized in the following eight points.

Sketch of the Algorithm

1. Input the values of the nasal airflow signal measured at regular intervals of 1/16th of a second from time 0 up to time T.
2. Normalize the signal. This is done by rescaling the original signal so that the total volume of inspiratory flow over the first 600 s coincides with a preset value.
3. Compute a first s8 wavelet shrinkage estimate of the normalized airflow signal and obtain a derivative estimate of the normalized airflow signal by computing a moving average of two consecutive first differences of the s8 wavelet shrinkage estimate divided by 1/16.
4. Compute a Haar wavelet shrinkage estimate and a second s8 wavelet shrinkage estimate of the normalized airflow signal. This second s8 wavelet shrinkage estimate is not as smooth as the one computed in step 3.

FIG. 3. Nasal airflow-related tracings for subjects 1 and 17. The top panels display the observed signals. The panels in rows 2, 3, and 4 display the wavelet shrinkage estimates of the normalized signals obtained using the Haar and the s8 bases. The bottom panels display estimates of the derivatives of the normalized signals obtained from the first s8 wavelet shrinkage estimate.

5. Identify the NF periods as those stretches of time where both the Haar wavelet shrinkage estimate and the derivative estimate are "small" in absolute value [i.e., `abs(Haar estimate) < wave.small` and `abs(derivative) < derivative.upper`, where `wave.small` and `derivative.upper` are set thresholds].

6. Identify as end points of the I periods those time instants when, after a period of sustained increasing flow (as assessed by the fact that the value of the estimated derivative, averaged over several time instants, exceeds a set threshold `derivative.lower`), either (a) the first `s8` wavelet shrinkage estimate crosses from negative to positive values, or (b) an NF period begins. Note that it is possible for both (a) and (b) to be true at instants of time that are in close proximity and some special care is needed to deal with these situations. In particular, we adopt the convention that an end point of an I period can occur at the beginning of an NF period but never within its body or at its end.

7. Identify the beginning of the I periods by backtracking from the end of the I periods until the signal becomes positive (in fact, "almost" positive, i.e., until the second `s8` wavelet shrinkage estimate exceeds a "small" negative threshold `wave.small.I`). Note that it is possible to backtrack past the end of an NF period. In such cases we adopt the convention that the beginning of the I period coincides with the end of the NF period.

8. Output a summary that, for each complete breathing cycle identified in the time interval $[0, T]$, indicates:
   a. The beginning of the I period (which coincides with the beginning of the breathing cycle).
   b. The beginning of the E period (which coincides with the end of the I period).
   c. The end of the breathing cycle (which coincides with the end of the E period).
   d. The beginning and end of all NF periods detected within the cycle.

The panels in rows two to five of Fig. 3 display the Haar and `s8` shrinkage estimates and the derivative estimates of the two normalized signals. The output obtained through an implementation of the algorithm in the statistical programming language S-Plus is summarized graphically in the top line panels: the solid vertical lines denote the beginning of inspiration, the dotted vertical lines denote the beginning of expiration, and the top horizontal segments extend over the time intervals that have been identified as NF periods.

Performance on Simulated Signals

In this section we report on the results of some simulation studies that we conducted to assess how sensitive the performance of the algorithm is to varying input conditions and to the values of several tuning parameters.

A typical input flow signal used in the simulation was constructed by piecing together 1002 replicates of one of the three elementary building blocks depicted in Fig. 4 and by adding noise to the periodic curve thus obtained. Each building block constitutes an idealized realization of a typical breathing cycle 8 s long. The signal is zero at times 0, 3, and 8, is negative between 0 and 3, and is positive between 3 and 8. In other words, the I portion of the breath runs from time 0 to time 3 and the E portion runs from time 3 to time 8. Each building block has a unique minimum during the I period, where the flow level equals −4.5, and a unique maximum during the E period, where the flow level equals 4.5. In the final part of the E period, each building block slowly tapers off toward zero, thus mimicking an NF zone.

The three elementary building blocks are each made up of 10 smoothly joined piecewise quadratic portions and one linear portion. The linear portion is located in the region where the flow crosses over from negative to positive values. The first derivatives of the 11 portions match at all juncture points. In addition, the derivative from the right is zero at time 0 and the derivative from the left is zero at time 8. Hence, when the individual building blocks are pieced together to form a signal to be used as input for the simulation study, they connect smoothly (with a horizontal tangent) at their juncture points.

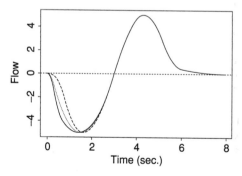

FIG. 4. The three elementary building blocks used to construct the input signals for the simulations.

The main purpose of the simulation study was to understand how various characteristics of the signal affect the accuracy with which the algorithm can identify the beginning of the I periods and the beginning of the E periods and how, by tuning some of the parameters of the algorithm, its performance can be improved. Our experience with the signals of the 20 subjects suggested that the algorithm can typically identify the beginning of the E periods with a high level of accuracy. This is primarily due to the fact that signals tend to cross rapidly from negative to positive values. Identification of the beginning of the I periods is more complicated because of the presence of the NF periods at the end of the E periods. In particular, the ability of the algorithm to detect reliably the beginning of an I period is greatly influenced by how rapidly inspiration resumes after the end of an E period and by the level of noise with which the signal is recorded.

With this in mind, we constructed the three elementary building blocks of Fig. 4 so that they drop off more or less rapidly after time 0. We will refer to the three different drop-off rates as rates 1, 2, and 3, with 1 denoting the most rapid rate of decrease (the solid line in Fig. 4), 2 denoting the intermediate rate of decrease (the dotted line in Fig. 4), and 3 denoting the slowest rate of decrease (the dashed line in Fig. 4). The latter part of the signal is the same for all three building blocks. In particular, the quite rapid rate of upcrossing at time 3 is the same for all three building blocks. The results of simulation studies not reported here confirmed our preliminary observation that the performance of the algorithm in the detection of the beginning of the E periods was unaffected by choosing such a rate within a range of fairly fast values typical of the signal for the subjects under study.

Another important factor affecting the performance of the algorithm is the noise level in the recorded signal. To quantify the impact of noise on the detection of the beginning of the I periods we perturbed each input signal used in the simulation with additive Gaussian noise. We used four different values for the standard deviation of the perturbation to be added at each sampled point (noise SD = 0.01, 0.05, 0.1, and 0.3). Each of the four panels in Fig. 5 depicts four breath cycles obtained by piecing together four basic building blocks with drop-off rate 3 and added Gaussian noise. The noise level is smallest in the top panel and largest in the bottom panel.

There are several parameters that can be tuned to optimize the performance of our algorithm. Among these is the `smooth.levels` argument in the S-Plus function `waveshrink()` described previously. For the Haar and the s8 shrinkage estimates of the flow signal computed in step 4 of the algorithm we considered setting the value of `smooth.levels` equal to 1, 2, and 3. This gave rise to nine possible categories of smoothing corresponding to all possible paired combinations of the three levels of smoothing for

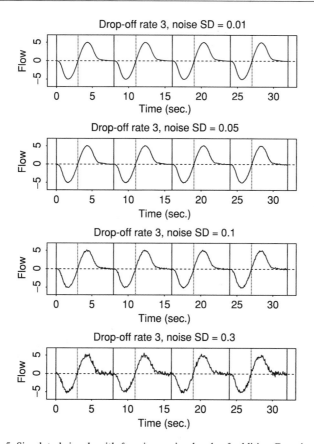

FIG. 5. Simulated signals with four increasing levels of additive Gaussian noise.

the Haar and the s8 estimates. We labeled the nine smoothing categories consecutively according to the scheme summarized in Table I.

Other important parameters that affect the performance of our algorithm are the various threshold levels to which the sizes of the various signal and derivative estimates are compared. In all simulation runs we set derivative.upper (defined in step 5) equal to $(0.05) \times f$ and derivative.lower (defined in step 6) equal to $(0.01) \times f$, where $f = 16$ is the sampling frequency. We considered four combinations of values of wave.small and wave.small.I (defined in steps 5 and 7), yielding the four thresholding categories summarized in Table II.

The simulation was set up to understand the interplay between important characteristics of the signal and the settings of critical tuning parameters of the algorithm. As explained previously, we controlled 3×4

TABLE I
LABELING OF THE SMOOTHING CATEGORIES

Smoothing category	1	2	3	4	5	6	7	8	9
smooth.levels value for the Haar estimate	1	1	1	2	2	2	3	3	3
smooth.levels value for the s8 estimate	1	2	3	1	2	3	1	2	3

TABLE II
LABELING OF THE THRESHOLDING CATEGORIES

Thresholding category	1	2	3	4
Value of wave.small	0.1	0.05	0.05	0.01
Value of wave.small.I	0.025	0.025	0.01	0.01

characteristics of the signal (drop-off rate × noise) and 9 × 4 tuning parameters of the algorithm (smoothing × thresholding), for a total of 3 × 4 × 9 × 4 = 432 experimental conditions. For each of the 432 experimental conditions we generated an input signal by adjoining 1002 building blocks with the appropriate drop-off rate and by perturbing it with the appropriate additive noise. To reduce variability not directly attributable to the experimental conditions, the same sequence of additive noise terms (appropriately rescaled to account for the different standard deviations) was used to generate all 432 input signals.

We ran the algorithm on each of the 432 input signals with the smoothing and thresholding tuning parameters set at the appropriate levels for the given experimental condition. We focused on the algorithm's performance in identifying the I periods. To discount possible edge effects, we eliminated the first and the last cycles from consideration. For a given signal, there were thus 1000 possible starting points of an I period that could be correctly identified by the algorithm. We denoted by I_c the set of $n_c \leq 1000$ output time points $\hat{t}_j$ correctly identified by the algorithm as the beginning of I periods. Essentially, we considered as a correct identification any output time point that was close to the true beginning of an I period and for which no other output time point was closer.

Besides producing correct identifications, the algorithm can make two types of errors, either by producing spurious identifications or by failing to produce correct identifications. These errors, which occur more frequently with highly noisy signals, are illustrated in Fig. 6. The starting point

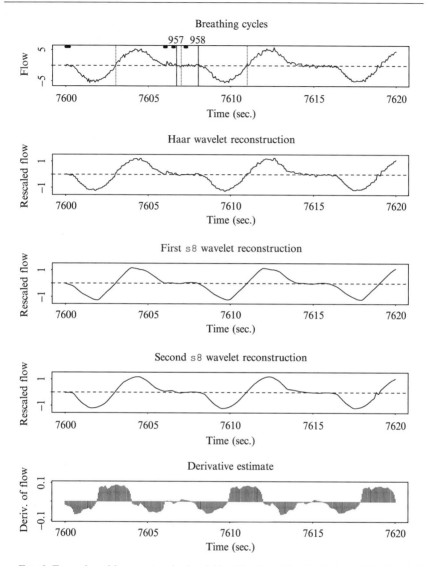

FIG. 6. Examples of incorrect and missed identifications. The beginning of the I period labeled 957 is a spurious identification. The beginning of the I period occurring at time 7616 was not identified.

of the I period labeled 957 is spurious. Because of the high noise level, step 6 of the algorithm was "tricked" into identifying as the end point of an I period (the dotted line within cycle 957) an upward crossing that was, in reality, a minor bump in the signal. In turn, by backtracking from the dotted line, step 7 of the algorithm made the spurious identification. Essentially, we considered as a spurious identification any output time point close to the true beginning of an I period for which there is another output time point that is closer.

Figure 6 also illustrates a situation in which the beginning of an I period (occurring at time 7616) was not identified. This happened as a consequence of the special handling of a rare pattern that is not explicitly described. Specifically, in step 6, the algorithm initially correctly identified the end point of an I period as occurring around time 7619. Yet, after backtracking during step 7, the algorithm identified incorrectly the starting point of an I period immediately to the left of time 7619. The reason for this becomes apparent if one considers that the second s8 wavelet reconstruction in Fig. 6 exhibits multiple zero crossings in close proximity of time 7619. As part of the special handling of unusual patterns, the algorithm reviews all identified I periods and drops as likely artifacts those that are shorter than a small threshold. This resulted in the elimination of the short I period identified near time 7619. Essentially, we considered as a missing identification any case in which the algorithm did not yield an output time point in the vicinity of the true beginning of an I period.

For the n_c time points $\hat{t}_j$ in I_c, i.e., for the time points correctly identified as the beginning of I periods, we quantified the performance of the algorithm on the basis of the mean squared error (MSE) in detection accuracy defined by

$$\frac{1}{n_c} \sum_{\hat{t}_j \in I_c} \left(\hat{t}_j - t_j \right)^2$$

where t_j denotes the true occurrence time estimated by $\hat{t}_j$. Note that this is a *conditional* measure of accuracy, given that correct detections have been made.

We made no attempt to adjust the measure of accuracy for the missed or spurious identifications. Rather, we tallied the two types of errors separately. We decided to follow this approach because we considered the relative costs of the two types of error to be application dependent and thus impossible to incorporate into a global measure of detection accuracy. Under most experimental conditions the simulations produced few or no errors. Missed identifications occurred under all experimental conditions involving the highest noise setting and under some experimental conditions involving the second highest noise setting and thresholding category 1. Spurious identifications occurred only under experimental conditions involving the highest noise setting and thresholding category 4.

The MSE in detection accuracy under the various experimental conditions is summarized in Fig. 7. The figure is made up of a 3×4 grid of 12 panels, with each row corresponding to one of the three drop-off rates and each column corresponding to one of the four different values for the standard deviation of the noise. In turn, each panel is made up of a 4×9 grid of 36 rectangular tiles, with each row corresponding to one of the four thresholding categories and each column corresponding to one of the nine smoothing categories. Each rectangular tile is color coded according to the color bar on the right-hand side of the figure to represent the estimated MSE in detection accuracy for the experimental condition corresponding to the given tile.

Several features of the performance of the algorithm become apparent from an analysis of the figure. First, the MSE in detection accuracy is quite small for both drop-off rates 1 and 2, regardless of the noise level and of the settings of the various tuning parameters. The maximum values of the MSE in detection accuracy are 0.025 for drop-off rate 1 and 0.047 for drop-off rate 2. There are, of course, small variations in the performance at the different settings of the tuning parameters that cannot be detected from the picture. A more accurate analysis shows that setting the smoothing category equal to 3 and the thresholding category equal to 2 or 3 always yields a performance that is optimal or close to optimal.

At drop-off rate 3 the thresholding category plays a more important role. Clearly, Fig. 7 suggests that setting the thresholding category at level 4 optimizes performance. While this is true with regard to the MSE in detection accuracy one must be cognizant of the missed and spurious detections that might occur. For example, all experimental conditions involving the highest noise level and thresholding category 4 yielded at least one spurious detection and as many as nine. At different thresholding levels there were no spurious identifications, but, under certain experimental conditions, the number of missed identifications was substantial. For example, under the highest noise level and thresholding category 1, the largest number of misses for the various experimental conditions turned out to be 139. There were, however, substantially fewer misses for the other thresholding categories. With the noise level held at its highest setting, there were no more than 17 misses at thresholding categories 2 and 3 and no more than 2 at thresholding category 4. Thus, if one uses a global measure of performance accounting not only for the MSE in detection accuracy but also for the missed and spurious identifications, then, as before, smoothing category 3 and thresholding category 2 or 3 yield compromise choices that guarantee quite reliable performances. Clearly, smoothing categories 7, 8, and 9 should be avoided.

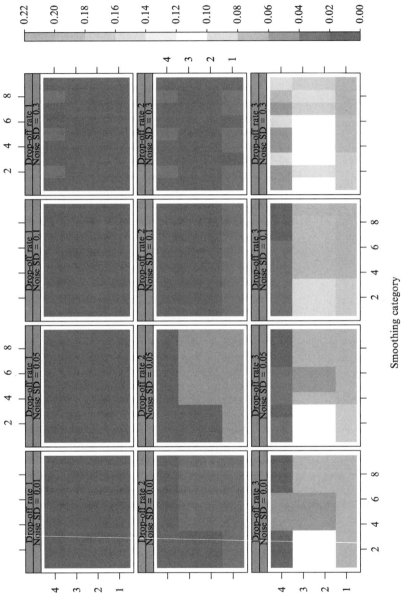

FIG. 7. Estimated MSE in detection accuracy. (See color insert.)

There are two additive components to the MSE in detection accuracy, the variance of the estimator and the square of the bias. In many cases the latter can be substantial, accounting for a large portion of the MSE. In practical applications it is often important to assess not only the size of the bias, but also its sign. Using plotting conventions similar to those used in Fig. 7, Fig. 8 illustrates the behavior of the bias under the various experimental conditions. The most important feature that becomes apparent from the figure is that under the vast majority of experimental conditions, the bias was estimated to be positive (purple to blue shades). This means that there tends to be a delay in the estimation of the occurrence times of the beginnings of the I periods.

The most important determinant of the bias appears to be the thresholding category, which affects the amount of backtracking in step 7 of the algorithm, both directly through the value of wave.small.I and indirectly through the value of wave.small, which in turn determines what is identified as an NF period in step 5. Note that at drop-off rate 3, the expected delay at threshold category 1 can be quite large (of the order of half a second). The only negative values of the bias (yellow and orange shades) were estimated under certain experimental conditions with the thresholding category set at 4. This happened especially under high noise levels that favored backtracking past the true beginnings of the I periods in step 7 of the algorithm. The previously recommended combinations with the smoothing category set at 3 and the thresholding category set at 2 or 3 appear to produce acceptable performances also in terms of the estimated bias.

An Example of a Derived Measure of Breathing

As we noted in the introduction, the analysis of airflow data performed by laboratory technicians yields a categorical classification of breathing episodes. Software programs for automated analysis that are currently available tend to mimic the process performed by laboratory technicians (see, for example, the Sandman Elite digital sleep software system from Mallinckrodt Inc, St Louis, www.mallinckrodt.com). We, on the other hand, are developing an automated system for the analysis of polysomnography breathing data that is based on the observation that breathing in general, and breathing during sleep in particular, is a continuous activity that is best described by continuous measures. In two studies[9,10] we demonstrated that

---

[9] M. Mullins, M. Johnson, M. Peruggia, and P. M. Suratt, Quantitating sleep disordered breathing, looking beyond the apnea hypopnea index (AHI). Am. Thoracic Soc. Annual Meetings, 2001.

[10] M. Mullins, M. Johnson, M. Peruggia, and P. M. Suratt, Duty cycle prolongation in obstructive sleep apnea, Am. Thoracic Soc. Annual Meetings, 2001.

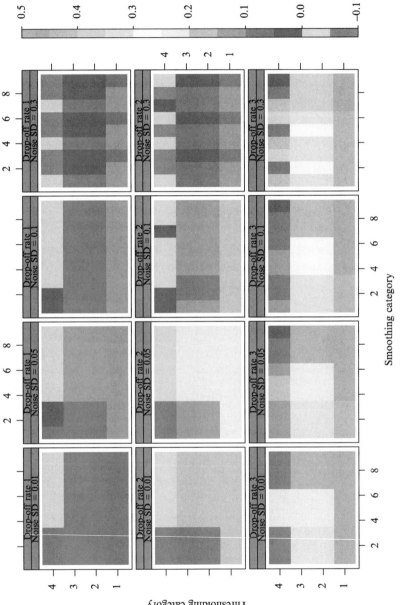

FIG. 8. Estimated bias in detection accuracy. (See color insert.)

it is possible to go beyond a discrete quantitation of sleep-disordered breathing by showing that continuous measures of sleep are more variable in diseased than in normal subjects.

The information about the ventilatory volume supplied by the nasal airflow signal is an important element for the detection and diagnosis of sleep-disordered breathing, but it may be imprecise due to the lack of calibration of the pressure transducer used to record it and/or to the possibility that the patient breathes through the mouth. Our automated algorithm for processing the airflow signal provides output related to the timing of certain events that occur while the patient is undergoing the sleep study (the I/E/NF periods) that are fairly insensitive to lack of calibration and to the presence of mouth breathing.

In this section we suggest how further processing and analysis of these variables will aid understanding of the nature of breathing during sleep on a continuous rather than categorical basis. To illustrate this point, we consider the NF periods identified by the algorithm. The NF periods are events in the breathing cycle that both laboratory technicians and currently available software programs fail to identify. Yet, they provide substantial information with regard to the nature of the airflow signal that might be utilized for diagnostic purpose.

Figure 9 helps to clarify the point we just made. The figure displays two histograms based on data for the two subjects whose nasal airflow traces are shown in Fig. 3. Each histogram summarizes the empirical distribution of the proportional duration of the NF periods within each of the breathing cycles occurring over an interval of 600 s. In other words, for each subject, all complete breathing cycles occurring in the given 600-s interval were first identified by means of the automated algorithm. Next, the proportional duration of the NF periods within each cycle was evaluated as the ratio, Tnf.Ttot = Tnf/Ttot, of the total length of the NF periods, Tnf, to the total length on the breathing cycle, Ttot. Finally, the histogram of these proportional durations was constructed.

Visual inspection of the two histograms reveals obvious differences in the distributions of Tnf.Ttot for the two subjects. The distribution for Subject 1, who was eventually diagnosed as being normal, is roughly symmetric about the median value of 0.21. Subject 17, who was eventually diagnosed as having severe sleep-disordered breathing, is different. The Tnf.Ttot distribution for Subject 17 spans almost the entire range of possible values and can be described roughly as a mixture of three components. In fact, in addition to a group of values that is typical of normal breathing, it is possible to observe a group of larger values that is typical of sleep cycles characterized by apnea episodes. Furthermore, there is a spike in the histogram corresponding to values around zero. This reflects the presence of recovery

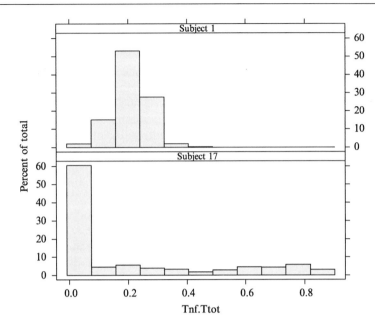

Fig. 9. Empirical distributions of the proportional durations of the NF periods for two subjects.

breaths occurring after apnea episodes that are characterized by an increased cycle frequency without intervening ventilatory pauses.

The existence of differences in the empirical distributions of Tnf.Ttot underscores the importance of our novel analysis of the NF periods, because it suggests that it is possible to recognize different types and stages of illness on the basis of properties of the distribution itself. However, due to the complicated nature of these differences, the properties of the distribution that need to be recognized cannot be easily described and summarized by one or two numbers such as the sample mean and standard deviation. Clearly, to be effective, a diagnostic tool would have to incorporate a more comprehensive portrayal of the distribution and this is the topic of ongoing research.

Discussion

The vast majority of the scientific literature on wavelet signal estimation hinges on the premise that the ultimate goal of the analysis is to obtain a reconstruction of the signal that minimizes the integrated mean squared error of estimation. The latter is an average measure of agreement of the estimate with the true value of the signal, and is appropriate only if what is needed is an approximation that is good on average.

Our conceptual contribution to the theory of signal estimation is that the integrated mean squared error criterion, while a popular one, should by no means be the only measure of performance to be employed. In our case, special features of the signal were of primary interest (namely, the end points of the I/E/NF periods) and our performance criterion, the MSE in detection accuracy of the beginning of the I periods, implicitly recognizes this fact.

Note that the s8 wavelet shrinkage estimate would provide a more than adequate average representation for the signals. Yet, by itself, it could not be used to detect reliably the end points of the I/E/NF periods. The main obstacle is that in zones of low flow, the derivative estimate based on the s8 wavelet shrinkage estimate is quite sensitive to artifacts due to noise and to the presence of little bumps in the flow. Thus, when at first we tried to identify the NF periods as those zones where the absolute value of the s8 wavelet shrinkage estimate was small and the absolute value of the derivative estimate was below some threshold, L, we faced a dilemma. By setting L at a small value we would end up detecting numerous spurious breaks within the NF periods in connection with the artifacts and the bumps. Trying to alleviate this problem by increasing the value of L would, unfortunately, result in the identification of several spurious NF periods.

The introduction of the Haar wavelet shrinkage estimate eliminated this conundrum. The algorithm now identifies the NF period essentially on the basis of the fact that Haar wavelet shrinkage estimate is small in absolute value over a certain region. The threshold value for the derivative, derivative.upper, mentioned in step 5 of the algorithm can be set fairly large, so that it plays only a minor role and the detection of the NF periods is not affected by the presence of artifacts and bumps. The reason why we maintained a condition on the size of the derivative at all is to avoid identifying as NF periods small regions around points of genuine zero crossing where the Haar representation of the signal is small in absolute value. Note that especially with highly noisy signals, an adequate tuning of the smoothing level of the Haar wavelet shrinkage estimate computed in step 4 of the algorithm is important because it will lessen the impact of artifacts like the one that led to the spurious detection in Fig. 6.

In summary, from a practical standpoint, we have presented a reliable algorithm for the analysis of nasal airflow traces. From a more conceptual standpoint, we have elaborated two main ideas that will contribute to further developments in the area of nonparametric curve estimation and signal analysis. First, we illustrated how application-specific measures of performance that are based on the features of interest of the signal can be derived and should be preferred to integrated mean squared error as the criterion for evaluating the quality of an estimation technique. Second, our algorithm makes use of the estimate of the derivative of the signal. The estimation of a signal derivative is notoriously a difficult problem. The strength of our approach is to understand how the reliability of the estimate that we compute is not uniform over time and to recognize the need to supplement and complement the information supplied by the estimate over those regions where it is less reliable.

Acknowledgments

This material is based upon work supported by the National Institutes of Health under Grants HL62401 and MO1RR00847 and by the National Science Foundation under Grant SES-0214574 and Agreement DMS-0112050. Junfeng Sun's work on this project was supported by an Ohio State Department of Statistics Graduate Research Associateship Award. Mario Peruggia would like to thank Peter Craigmile and Steven MacEachern for many helpful discussions.

[9] Quantifying Asynchronous Breathing

By MICHAEL L. JOHNSON and PAUL SURATT

Introduction

Respiratory inductance plethysmography is reported to be a sensitive method of detecting thoracic and abdominal asynchrony in children with obstructive sleep-disordered breathing (OSDB). In practice, asynchrony is usually evaluated subjectively and not quantified. A reader visually reviews the entire sleep record and makes an assessment as to whether asynchrony occurred and, if possible to determine, whether it was more common in certain sleep stages, body positions, or disordered breathing events such as apneas and hypopneas. While this method has allowed us to understand much about OSDB in children, it is labor intensive and imprecise. In children in whom it has been stated that more than one apnea or hypopnea per hour of sleep is abnormal, precision in determining thoracic and abdominal asynchrony (i.e., paradoxical breathing) is critically important.

A number of studies have shown it is possible to quantify thoracic and abdominal synchrony and asynchrony using several different algorithms. This quantitative analysis is rarely used in clinical or research polysomnographic studies of children, perhaps in part because it is difficult to integrate the quantitative analysis with other portions of the study, i.e., what stages of sleep and sleeping positions it occurred in, and whether it was associated with apneas and hypopneas.

We describe a simple computer algorithm of quantifying inductance plethysmography thoracic and abdominal signals and integrating it with sleep stages, disordered breathing events, and body position that can be used in clinical and research polysomnographic studies. The method of quantifying asynchrony is a modification of the linear transfer function method, a method that has been shown to be extremely accurate.

Methods

The tradition method of analysis requires a technician to visually review the entire sleep record and make an assessment as to whether asynchrony occurred and, if possible to determine, whether it was more common in certain sleep stages, body positions, or disordered breathing events such as apneas and hypopneas. The difficulty is that a typical 8-h overnight polysomnography sleep record consists of approximately 10,000 breaths with abnormalities occurring a few times per hour. These breathing abnormalities have durations as short as 6 s. Thus, it is both time consuming and the manual assessment is likely to be inaccurate.

We tested this method in children taking part in a UVA IRB-approved study of OSDB. Children with adenotonsillar hypertrophy who were (1) thought to have OSDB or (2) thought not to have OSDB (controls) underwent overnight polysomnography. Those with OSDB and adenotonsillar hypertrophy were reevaluated 6–12 months after removal of their tonsils and adenoids; controls were also reevaluated 6–12 months later.

Overnight polysomnography consisted of monitoring with inductance plethysmography (Ambulatory monitoring, Ardley, NY), nasal pressure detected by nasal pressure, pulse oximetry, ECG, as well as electroencephalograms, C3O2 and O2A1, electrooculograms, and submental electromyograms. These are typically digitized at frequencies from 4 to 256 Hz and thus the entire data set can easily require 50 megabytes of storage.

When the subject is breathing normally, thoracic and abdomen movements are almost perfectly synchronous and simultaneous. When there is OSDB, thoracic and abdomen movements may become asynchronous or paradoxical. Consequently, the identification of OSDB involves processing large amounts of data looking for infrequent paradoxical events of very short duration.

Our approach involves segmenting the thoracic and abdomen extension time series into short intervals of typically 5-s durations. The 5-s duration is somewhat arbitrary; it is shorter than the expected event duration, it is

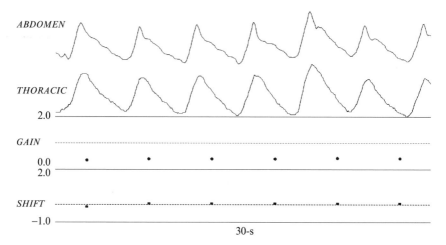

FIG. 1. A typical 30-s epoch of nonparadoxical breathing as measured by abdominal and thoracic extension. The numbers above and below the labels on the left are the ranges of the data. The units of the abdominal and thoracic signals are directly from the polysomnographic recording and have not been calibrated. Note that the shift corresponding to each of the 5-s segments is not significantly different from zero. The *GAIN* for each of the segments is approximately 0.35, indicating that the amplitude of the thoracic signal is approximately 35% of the abdominal signal. This patient exhibited obstructive sleep-disordered breathing. This is a selected epoch (sleep stage 2) that did not display a large amount of paradoxical breathing.

longer than a typical breath, and it is an integer divisor of the standard 30-s epoch traditionally utilized in polysomnography studies. Other durations such as 4, 6, and 10 s should also work with this algorithm since it is not specific to the 5-s window. A typical 30-s epoch of nonparadoxical breathing is shown in Fig. 1.

For each of the short (e.g., 5 s) segments of the thoracic and abdomen extension time series we assume that the thoracic time series can be described as a time shifted linear transfer function of the abdomen time series, i.e.:

$$THORACIC[time] = GAIN * ABDOMEN\ [time + SHIFT] + OFFSET \tag{1}$$

For each of the segments we determine three quantities: *GAIN, OFFSET,* and *SHIFT.* If the breathing is normal then *SHIFT* is expected to be nearly zero and if the breathing is paradoxical then *SHIFT* is expected to have a positive nonzero value. The *GAIN* and *OFFSET* are related to the exact placement and sensitivity of the sensors on the subject, in addition to the paradoxical nature of the breathing, and are thus less useful in identifying paradoxical breathing. It is their relative timing that is important.

The *GAIN, OFFSET,* and *SHIFT* are determined for each of the segments by testing all possible *SHIFT* values from −1.0 to 1.6 s. For each of the potential shifts the *THORACIC* signal is shifted and the *GAIN* and *OFFSET* are determined by a linear least-squares parameter estimation procedure according to Eq. (1). If the thoracic and abdomen signals are recorded at 16 Hz then this corresponds to 42 individual parameter estimations. For each of the possible shifts the variance of fit is evaluated and the optimal *SHIFT* is evaluated as the shift that corresponds to the lowest variance of fit that corresponded to a positive *GAIN*. A graph of the normalized variance of fit as a function of the shift for the second segment in Fig. 1 is presented in Fig. 2 as an example of this procedure.

Note that for this procedure the shifted time series are not truncated to fit within the individual 5-s segments. Thus, if the abdominal signal is shifted by +1.6 s then the thoracic signal for the current segment is fit as a function of an abdominal signal that starts with the last 1.6 s of the previous segment and ends 1.6 s before the end of the current segment.

The positive *GAIN* constraint is imposed because these are repeating signals. The consequence of applying this algorithm to a repeating signal can be easily seen by considering a pair of periodic sine or cosine waves. When they are perfectly aligned, the variance of fit will be small and the *GAIN* will be positive. However, if they are 180° out of phase, they will also have a small variance of fit but the *GAIN* will be negative. Thus, the positive *GAIN* is required to preclude an out-of-phase signal.

The allowed *SHIFT* range −1.0 to 1.6 s was chosen because it represents a typical range of expected paradoxical shifts without extending into the previous or next breath. Using the sine or cosine wave example the allowed *SHIFT* range was chosen to preclude being ±360° out of phase.

An estimate of the standard error of the shift can be obtained by applying an F test for an additional parameter.[1] For the present example, an increase in the variance of fit of approximately 2.5% above the optimal (minimal) value corresponds to a $p < 0.05$. Thus, the estimated value of the *SHIFT* is extremely well determined for the example shown in Fig. 2. Estimates of the standard errors of the *GAIN* and the *OFFSET* can be obtained by standard linear model least-squares analysis.

Note that the estimated uncertainties of the *GAIN, OFFSET,* and *SHIFT* are only approximate because they consider only the experimental

[1] P. R. Bevington, *in* "Data Reduction and Error Analysis for the Physical Sciences," p. 203. McGraw-Hill, New York, 1969.

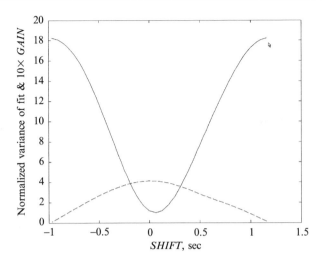

Fig. 2. The normalized variance of fit (solid line) as a function of the shift for the second 5-s segment in Fig. 1. Each estimated variance of fit is normalized by dividing by the lowest variance of fit that corresponded to a positive *GAIN*. The dashed line corresponds to 10 times the *GAIN*.

uncertainties of the thoracic signal. The experimental uncertainties of the abdominal signal are ignored. Thus, the uncertainties of the *GAIN,* *OFFSET,* and *SHIFT* are actually underestimates of the true uncertainties. They are, however, useful when considering relative errors of different 5-s segments.

Results

Figures 1 and 3 present a selected epoch of a patient with OSDB. Figure 1 displays the data from a sleep stage 2 epoch where the subject does not display a significant amount of paradoxical breathing. Figure 3 presents a REM sleep stage epoch from the same subject that exhibits a significant amount of paradoxical breathing.

Figure 4 presents the corresponding nonparadoxical breathing for a normal control subject. Note that Figs. 1 and 4 both correspond to nonparadoxical breathing but the waveforms for the breathing are quite different. The waveforms are more consistent between Figs. 1 and 3, which show the obstructive sleep-disordered breathing patient during an epoch with nonparadoxical breathing and while breathing paradoxically.

Examining single epochs of data does not address the problem of a technician having to examine approximately 2000 individual epochs for

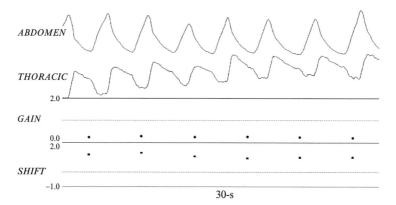

Fɪɢ. 3. A REM sleep stage epoch from the same obstructive sleep-disordered breathing patient as in Fig. 1 that displays a significant amount of paradoxical breathing.

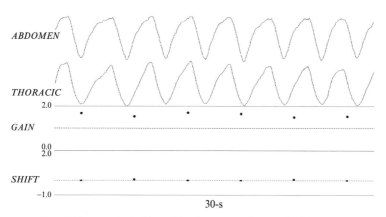

Fɪɢ. 4. Nonparadoxical breathing for a normal control subject.

each sleep study visit. Figure 5 presents an example of *SHIFT* values from a 6.4-h sleep record of a control subject. Figure 5A presents the values calculated for the entire time. Note that there are many values that approach either −1.0 or + 1.6 s. Figure 5B presents the same data except that the technician scoring of the sleep stages has been used to eliminate areas where the subject was awake. Areas where the data recording system failed have also been removed. It is clear from a comparison of Fig. 5A and 5B that the large *SHIFT* values can be used to predict areas where the subject is awake.

Two other areas of Fig. 5B are particularly notable. The large *SHIFT* value labeled **b** is the first 5-s segment of a technician score sleep epoch

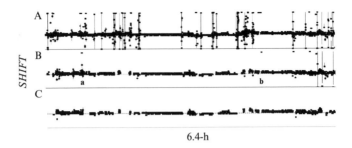

6.4-h

FIG. 5. An example of *SHIFT* values of a 6.4-h sleep record of a normal volunteer. (A) The values calculated for the entire time. (B) The same results except that the technician scoring of the sleep stages based upon the EEG data has been used to eliminate areas where the subject was awake. (C) The same series of *SHIFT* values where wake epochs, bad data epochs, arousal epochs, and sleep epochs at the edge of a wake epoch have been removed.

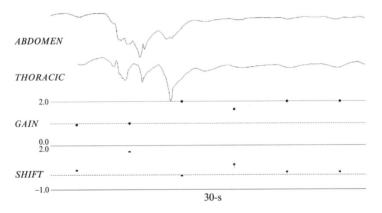

FIG. 6. An expanded view of the epoch labeled **a** in Fig. 5.

immediately following a technician score wake epoch. Note that the technician will score an epoch as a particular sleep stage if the majority of the time in the 30-s epoch is of that sleep stage, even if 14 s is awake or in a different stage. Several of the other outlier points in Fig. 5B correspond to such edge effects. The large *SHIFT* values in the area labeled **a** correspond to an area marked by the technician as an 8-s-long arousal starting 4 s into the epoch. Figure 6 presents an expanded view of this epoch that clearly indicates the period of the arousal as an abnormal area. Figure 5C presents the same series of *SHIFT* values where wake epochs, bad data epochs, arousal epochs, and sleep epochs at the edge of a wake epoch have been removed.

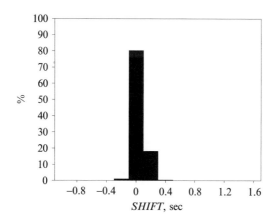

Fig. 7. A histogram of the *SHIFT* values presented in Fig. 5C.

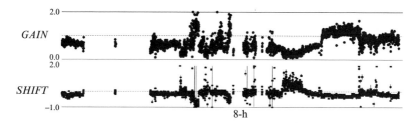

Fig. 8. The *SHIFT* and *GAIN* results from an 8-h sleep study of a patient with obstructive sleep-disordered breathing. The technician scoring of the sleep stages based upon the EEG data has been used to eliminate areas where the subject was awake, as well as bad data regions.

A histogram of the *SHIFT* values presented in Fig. 5C is presented in Fig. 7. The distribution characteristics were a mean of +0.0381 s with a 95% confidence region of 0.0356–0.0406 s. The standard deviation was 0.0686. The variance was 0.0472 with a 95% confidence interval of 0.0448–0.0497 and 2922 values. The skewness was 0.2733 and the kurtosis was 0.6052.

To demonstrate that this procedure has potential for investigating obstructive sleep-disordered breathing, the *GAIN* and *SHIFT* values from an 8-h sleep study of a patient with obstructive sleep-disordered breathing are shown in Fig. 8. For comparison, Fig. 9 presents the same subject evaluated approximately 9 months after the removal of their tonsils and adenoids. Clearly, the before study has a much broader distribution of *SHIFT* values than does the corresponding study after removal of their tonsils and

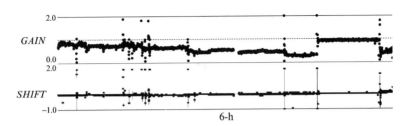

Fig. 9. The *SHIFT* and *GAIN* results from a 6-h sleep study of the same patient as shown in Fig. 8 approximately 9 months after removal of their tonsils and adenoids. The technician scoring of the sleep stages based upon the EEG data has been used to eliminate areas where the subject was awake, as well as bad data regions.

adenoids. The before standard deviation (Fig. 8) was 0.2608 and the after standard deviation (Fig. 9) was 0.1194.

Also notice in Figs. 8 and 9 that as in Fig. 5B, the extreme values of *SHIFT* values mark positions of particular interest such as arousals and awake intervals. This can also be seen as discontinuities in the *GAIN* signals that are also presented in Figs. 8 and 9.

Conclusion

We describe a robust computerized algorithm of detecting and describing asynchronous movements of the chest and abdomen during sleep. This algorithm can be performed rapidly with a laptop computer and can be easily integrated into results from polysomography.

The studies presented, and others that have not been presented here, indicate that this algorithm may be useful for the diagnosis of obstructive sleep-disordered breathing. They also indicate that it may also be of use to speed technician scoring polysomnography studies by indicating regions of the study that contain events of interest such as arousals and awake periods.

This algorithm is particularly useful since the measurements of chest and abdomen movements are nonintrusive. Thus, it can potentially be used in an outpatient environment and with children.

Acknowledgments

The authors acknowledge the support of the National Science Foundation Science and Technology Center for Biological Timing at the University of Virginia (NSF DIR-8920162), the General Clinical Research Center at the University of Virginia (NIH RR-00847), the University of Maryland at Baltimore Center for Fluorescence Spectroscopy (NIH RR-08119), and NIH Grant HL62401.

[10] Mixed-Model Regression Analysis and Dealing with Interindividual Differences

By HANS P. A. VAN DONGEN, ERIK OLOFSEN,
DAVID F. DINGES, and GREG MAISLIN

Introduction

Repeated-measures study designs are popular in biomedical research because they allow investigation of changes over time within individuals. These temporal changes in subjects may themselves be of interest (e.g., to document the effects of aging), or they may enable statistically powerful comparisons of different conditions within the same individuals (e.g., in cross-over drug studies). Complications arise when analyzing the longitudinal data from repeated-measures studies involving multiple subjects, however, as the data from the different subjects must be combined somehow for efficient analysis of the within-subjects changes. After all, the temporal changes that the various subjects have in common are typically of greater interest than the temporal profiles of each individual subject. The complications stem from the fact that the data collected within a subject are not independent of each other—they are correlated (and the magnitude of this correlation is usually not a priori known).[1] The data from different subjects, on the other hand, are typically independent. To properly analyze data from an experiment that combines multiple subjects with multiple data points per subject, two distinct sources of variance in the overall data set must be considered: between-subjects variance and within-subjects variance. Statistical analysis techniques targeted specifically at longitudinal data must keep these sources of variance separated.[2]

In this chapter, we consider mixed-model regression analysis, which is a specific technique for analyzing longitudinal data that properly deals with within- and between-subjects variance. The term "mixed model" refers to the inclusion of both fixed effects, which are model components used to define systematic relationships such as overall changes over time and/or experimentally induced group differences; and random effects, which account for variability among subjects around the systematic relationships captured by the fixed effects. To illustrate how the mixed-model regression approach can help analyze longitudinal data with large interindividual

---

[1] Counter-intuitively, it is precisely this feature of repeated-measures experimental designs that makes them statistically powerful and efficient.[2]

[2] P. Burton, L. Gurrin, and P. Sly, *Stat. Med.* **17,** 1261 (1998).

differences, we consider psychomotor vigilance data from an experiment involving 88 h of total sleep deprivation, during which subjects received either sustained low-dose caffeine or placebo.[3,4] We first apply traditional repeated-measures analysis of variance (ANOVA), and show that this method is not robust against systematic interindividual variability. The data are then reanalyzed using linear mixed-model regression analysis in order to properly take into account the interindividual differences. We conclude with an application of nonlinear mixed-model regression analysis of the data at hand, to demonstrate the considerable potential of this relatively novel statistical approach. Throughout this chapter, we apply commonly used (scalar) mathematical notation and avoid matrix formulation, so as to provide relatively easy access to the underlying statistical methodology.

Experiment and Data

A total of $n = 26$ healthy adult males (moderate caffeine consumers) participated in a 10-day laboratory study. Following a 2-week period in which they refrained from caffeine use, subjects entered the laboratory. After one adaptation night, subjects had two baseline days with bedtimes from 23:30 until 07:30. They then underwent 88 h of total sleep deprivation, during which time they were constantly monitored and kept awake with mild social stimulation. The experiment concluded with three recovery days. Every 2 h of scheduled wakefulness, subjects were tested on a 30-min computerized neurobehavioral assessment battery, which included a 10-min psychomotor vigilance task (PVT). Psychomotor vigilance performance was assessed by counting the number of lapses, defined as reaction times equal to or greater than 500 ms, per 10-min test bout. For the purposes of the present analyses, overall daytime performance was determined by averaging the test bouts at 09:30, 11:30, 13:30, 15:30, 17:30, 19:30, and 21:30 for each day. This served to average out the natural circadian (24-h) rhythm in performance data.[5] The first daytime period during the 88 h of wakefulness, before any sleep loss was incurred, we call Day 0. The subsequent three daytime periods we refer to as Days 1, 2, and 3.

[3] D. F. Dinges, S. M. Doran, J. Mullington, H. P. A. Van Dongen, N. Price, S. Samuel, M. M. Carlin, J. W. Powell, M. M. Mallis, M. Martino, C. Brodnyan, N. Konowal, and M. P. Szuba, *Sleep* **23**(Suppl. 2), A20 (2000).
[4] H. P. A. Van Dongen, N. J. Price, J. M. Mullington, M. P. Szuba, S. C. Kapoor, and D. F. Dinges, *Sleep* **24**, 813 (2001).
[5] H. P. A. Van Dongen and D. F. Dinges, *in* "Principles and Practice of Sleep Medicine" (M. H. Kryger, T. Roth, and W. C. Dement, eds.), 3rd Ed., p. 391. W. B. Saunders, Philadelphia, PA, 2000.

Subjects were randomized to one of two conditions: $n_1 = 13$ subjects were randomized to receive sustained low-dose caffeine (0.3 mg/kg body weight, or about a quarter cup of coffee, each hour) and $n_2 = 13$ subjects were randomized to receive placebo, in double-blind fashion. Caffeine or placebo pill administration began at 05:30 after 22 h of sustained wakefulness, and continued at hourly intervals for the remaining 66 h of total sleep deprivation. At 1.5-h intervals on average, blood samples were taken via an indwelling intravenous catheter for assessment of blood plasma concentrations of caffeine[6] (these data were available for 10 subjects in the caffeine condition only).

The present investigation focuses on whether caffeine mitigated the psychomotor vigilance performance impairment resulting from total sleep deprivation, and if so, for how long. Figure 1 shows the psychomotor vigilance data (PVT lapses) for both conditions, as well as the caffeine concentrations in blood plasma for the subjects in the caffeine condition, during the 88 h of total sleep deprivation. The error bars (representing standard deviations) in this figure show large interindividual differences both in the plasma concentrations of caffeine and in the levels of psychomotor vigilance impairment, posing a challenge for the analysis of this data set. In addition, the random assignment to condition resulted in differences between the average performance levels for the two conditions even before the administration of caffeine or placebo. This is also evident in Fig. 2, which shows the daytime averages for psychomotor vigilance performance lapses for Day 0 (before pill administration) and across Days 1–3 (during pill administration). The data points used to construct Fig. 2 are given in Table I.

Repeated-Measures Analysis of Variance

As a first analysis of the psychomotor vigilance performance data for the two conditions across Days 0 through 3, we use a commonly applied technique called repeated-measures analysis of variance (ANOVA). Before describing repeated-measures ANOVA, we describe the simpler situation in which subjects in N independent groups each contribute only one observation to the overall data set, so that all data points are mutually independent. The data points within each group are assumed to be normally distributed with equal variance but possibly differing mean. If the variance within the groups is relatively small compared to the variance among the group means, then the differences among the means are significant, that is, larger than could have reasonably arisen merely from

---

[6] Caffeine concentrations were assessed by EMIT enzyme multiplication immunoassay (Syva, Palo Alto, CA).

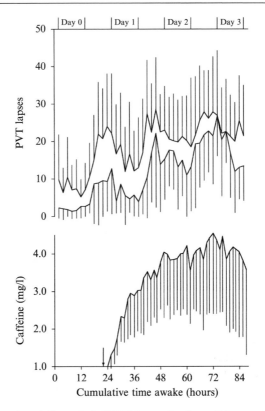

Fig. 1. Psychomotor vigilance task (PVT) lapses for the caffeine condition (downward error bars) and the placebo condition (upward error bars), and caffeine concentrations in blood plasma for the caffeine condition, across 88 h of total sleep deprivation. The curves represent condition averages; the error bars represent the standard deviation over subjects (suggesting large interindividual differences). The arrow indicates the beginning of hourly pill administration (caffeine or placebo).

variability in the data within the groups. This forms the basis of one-way ANOVA, as is illustrated in Fig. 3.

We define n as the total number of subjects, n_j as the number of subjects for group j ($j = 1, \ldots, N$), M_j as the observed mean for group j, and s_j^2 as the observed variance for group j:

$$s_j^2 = \{\Sigma_i (y_{ji} - M_j)^2\}/(n_j - 1) \tag{1}$$

where i identifies the different subjects in the group (say, $i = 1, \ldots, n_j$) and y_{ji} are the data points for these subjects. The variance components that form the basis of ANOVA are usually expressed in terms of sums of

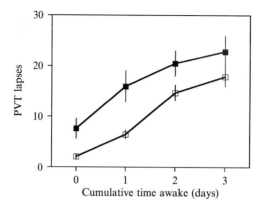

FIG. 2. Average daytime lapses on the psychomotor vigilance task (PVT) for the caffeine condition (open boxes) and the placebo condition (closed boxes) prior to pill administration (Day 0) and during pill administration (Days 1–3). The boxes represent condition means; the error bars represent standard errors of the mean.

squares (SS), such as the sum in curly brackets in Eq. (1), and in terms of mean squares (MS), defined as the sum of squares divided by the degrees of freedom (df). For the overall data set, which we indicate with S, the variance is thus described by

$$SS(S) = \Sigma_j\Sigma_i(y_{ji} - M)^2, \qquad df(S) = n - 1, \qquad MS(S) = SS(S)/df(S) \quad (2)$$

where M is the grand mean of the data.

The common within-groups variance, indicated here with w, is estimated using the group-specific variances:

$$SS(w) = \Sigma_j\Sigma_i(y_{ji} - M_j)^2, \qquad df(w) = \Sigma_j(n_j - 1) \quad (3)$$

so that

$$MS(w) = SS(w)/df(w) = \Sigma_j(n_j - 1)s_j^2/\Sigma_j(n_j - 1) \quad (4)$$

The between-groups variance (i.e., the variance among the group means) can then be computed from the difference between $SS(S)$ and $SS(w)$:

$$SS(b) = SS(S) - SS(w), \quad df(b) = N - 1, \quad MS(b) = SS(b)/df(b) \quad (5)$$

To evaluate the statistical significance of the difference between the group means (the "effect of condition"), the one-way ANOVA F statistic [with $df(b)$ and $df(w)$ degrees of freedom] is used[7]:

---

[7] When only two groups are compared [i.e., $df(b) = 1$], the square-root of the F statistic in Eq. (6) yields a t statistic for an equivalent t test with $df(w)$ degrees of freedom.

TABLE I
DAYTIME AVERAGE DATA IN TWO EXPERIMENTAL CONDITIONS[a]

id	cond	day0	day1	day2	day3
1	1	0.4	2.0	23.0	31.3
2	1	1.0	5.7	19.1	21.9
3	1	0.6	7.4	9.9	25.9
4	1	4.7	11.3	18.0	11.7
5	1	2.6	5.9	13.6	23.7
6	1	1.6	4.0	19.9	24.4
7	1	3.3	5.7	11.4	15.4
8	1	4.4	4.7	19.6	13.3
9	1	0.0	10.1	20.3	12.4
10	1	2.6	10.9	8.0	13.7
11	1	0.3	3.3	9.7	13.6
12	1	3.0	9.3	6.7	13.4
13	1	2.7	3.9	12.0	12.0
14	2	5.0	9.3	31.9	21.7
15	2	5.1	15.6	23.6	43.9
16	2	1.0	4.9	18.3	23.0
17	2	16.6	35.1	31.7	30.1
18	2	15.1	22.0	28.7	32.3
19	2	0.0	0.0	2.4	0.0
20	2	0.3	1.4	9.1	10.6
21	2	1.3	8.1	13.3	22.3
22	2	11.6	23.9	16.9	15.9
23	2	7.1	32.1	27.0	21.9
24	2	12.7	22.0	18.6	20.0
25	2	2.6	11.9	25.7	21.3
26	2	20.7	21.4	19.7	34.1

[a] Daytime averages are given for the number of psychomotor vigilance performance lapses per 10-min test bout, for each of the subjects (column "id") in the caffeine condition (column "cond" value 1) and the placebo condition (column "cond" value 2), across Days 0–3 of total sleep deprivation (columns "day0" through "day3").

$$F[\mathrm{df}(b), \mathrm{df}(w)] = \mathrm{MS}(b)/\mathrm{MS}(w) \qquad (6)$$

For the F test, it is assumed that the data are randomly sampled from normal distributions, although ANOVA results are robust to departures from this assumption. It is also assumed that the underlying distributions from which the data within each group are sampled have equal variances.

For the analysis of longitudinal data, the ANOVA method is adapted to compare subsets of the data measured in the same subjects at different time points, so as to test whether or not the mean for the subsets is the same at all time points. This is called repeated-measures ANOVA, and it differs from

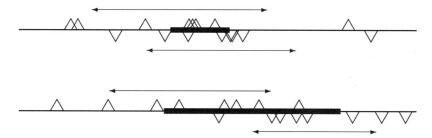

FIG. 3. Illustration of one-way analysis of variance (ANOVA) for two independent groups of eight data points each. On the upper line, data points in the first group (upward triangles) and the second group (downward triangles) show much overlap. The variances of the data in each of the groups, which are illustrated by the arrows (corresponding to the one-standard-deviation intervals around the group means), are relatively large compared to the variance between the group means, which is illustrated by the black bar (corresponding to the one-standard-deviation interval around the mean of the group means). Assuming normal distributions for the data, one-way ANOVA shows that these two groups are not significantly different ($F[1, 14] = 0.03, p = 0.96$). On the lower line, data points in the first group (upward triangles) and the second group (downward triangles) show less overlap. The variances of the data in each of these groups are relatively small compared to the variance between the group means. One-way ANOVA shows that these groups are significantly different ($F[1, 14] = 12.01, p = 0.004$).

one-way ANOVA in that individual data points can no longer be assumed to be independent. If there is only one experimental condition (i.e., $N = 1$), the variance in the data (in terms of sums of squares) is partitioned into two parts. One variance component is the variance among the means over time, which in repeated-measures ANOVA is a within-subjects factor since each subject is measured at all time points. The other variance component is residual variance, which is the remaining variance within individual subjects. The variance among the means over time is represented by

$$SS(A) = n\Sigma_t(m_t - M)^2, \qquad df(A) = T - 1 \qquad (7)$$

where t identifies the different time points ($t = 0, \ldots, T - 1$), and m_t is the mean at time t. The residual variance can be computed as

$$SS(R) = \Sigma_i\Sigma_t(t_{it} - m_t)^2 - SS(A), \qquad df(R) = (n - 1)(T - 1) \qquad (8)$$

where y_{it} are the data points for subject i at time t. The following F statistic is used to evaluate the statistical significance of the difference between the means of the different subsets (the "effect of time")[8]:

---

[8] The test for the effect of time is essentially a generalization of the paired-samples t test.

$$F[\mathrm{df}(A), \mathrm{df}(R)] = \mathrm{MS}(A)/\mathrm{MS}(R) \qquad (9)$$

Note that in this test for the effect of time, between-subjects variance (i.e., systematic differences among subjects over time) is automatically filtered out.

The principles of one-way ANOVA and repeated-measures ANOVA can be combined in a "mixed design" to compare independent conditions (groups) to each other over time. The full factorial form of the mixed design provides tests for the effect of time (applying to all conditions), for the effect of condition (applying to all time points), and for the interaction of condition by time. The interaction effect concerns any remaining systematic differences for the means among conditions and over time, after those applying to all conditions and those applying to all time points have been taken into account.[9] A mixed design involves partitioning the variance into within-subjects and between-subjects variance, with within-subjects variance represented by

$$\mathrm{SS}(W) = \Sigma_j \Sigma_i \Sigma_t (y_{jit} - M_{ji})^2 \qquad (10)$$

where i identifies different subjects in each condition j $(i = \sum_{j-1} n_j + 1, \ldots, \sum_j n_j)$, y_{jit} are the data points for these subjects at time points t, and M_{ji} is the mean over time for these subjects. The between-subjects variance can be shown to be the difference between the overall variance and the within-subjects variance (in terms of sums of squares), as follows:

$$\mathrm{SS}(B) = \mathrm{SS}(S) - \mathrm{SS}(W) \qquad (11)$$

For the purpose of testing the effect of condition, the between-subjects variance (in terms of sums of squares) is further partitioned into between-conditions variance (indicated by C) and error variance (indicated by E), as follows:

$$\mathrm{SS}(E) = T \ \Sigma_j \Sigma_i (M_{ji} - M_j)^2, \qquad \mathrm{df}(E) = \Sigma_j(n_j - 1) \qquad (12)$$

where M_j is the mean of all data for condition j, and

$$\mathrm{SS}(C) = \mathrm{SS}(B) - \mathrm{SS}(E), \qquad \mathrm{df}(C) = N - 1 \qquad (13)$$

The following F statistic is then used to evaluate the effect of condition:

$$F[\mathrm{df}(C), \mathrm{df}(E)] = \mathrm{MS}(C)/\mathrm{MS}(E) \qquad (14)$$

For the purpose of testing the effect of time and the interaction effect, the within-subjects variance is further partitioned into across-times variance (indicated by A) as given by Eq. (7), and interaction variance

[9] R. L. Rosnow and R. Rosenthal, *Psychol. Sci.* **6**, 3 (1995).

(indicated by I), and residual variance (indicated by R). The latter variance component is represented by

$$SS(R) = \Sigma_j\Sigma_i\Sigma_t[y_{jit} - (m_{jt} + M_{ji} - M_j)]^2, \quad df(R) = \Sigma_j(n_j - 1)(T - 1) \quad (15)$$

where m_{jt} is the mean for condition j at time t. The interaction is then computed as

$$SS(I) = SS(W) - SS(A) - SS(R), \qquad df(I) = (N - 1)(T - 1) \quad (16)$$

The following F statistic is used to evaluate the effect of time[10]:

$$F[df(A), df(R)] = MS(A)/MS(R) \quad (17)$$

Furthermore, the following F statistic is used to evaluate the interaction of condition by time:

$$F[df(I), df(R)] = MS(I)/MS(R) \quad (18)$$

We refer to the literature for information about the derivation of these formulas and for more in-depth discourses on repeated-measures ANOVA.[11]

Repeated-Measures Analysis of Variance Applied to the Data

Let us assume that we have the psychomotor vigilance performance data for Days 0–3 of the sleep deprivation experiment stored in a spreadsheet, organized in columns of $n = n_1 + n_2 = 26$ rows each: a column listing the different subjects by unique identifiers (named "id"), a column indicating in which condition each subject was (named "cond"), and four columns for the subjects' data from Day 0 until Day 3 (named "day0" through "day3," respectively). This conforms to the way the data are organized in Table I. To perform the calculations for repeated-measures ANOVA, we use the following general linear model (GLM) command in the computer software SPSS[12]:

```
GLM
    day0 day1 day2 day3 BY cond
    /WSDESIGN = time
    /WSFACTOR = time 4
    /DESIGN = cond.
```

---

[10] An adjustment must be made to SS(A) in the case of unequal numbers of subjects in the different conditions. The details of this adjustment for a "proportional sampling model" are omitted here.

[11] E. R. Girden, "ANOVA: Repeated Measures." Sage, Newbury Park, CA, 1992.

[12] SPSS for Windows, release 11.0.1. SPSS Inc., Chicago, IL, 2001.

The second line of this command identifies the data columns for the dependent variables, as well as the between-subjects factor column indicating to which conditions these data belong. The next two lines state that the dependent variables constitute repeated measures of a single underlying variable "time" with four levels (i.e., four different time points). The last line assigns the between-subjects factor (i.e., condition), making this a mixed-design repeated-measures ANOVA.

Table II shows the results of this analysis, revealing a significant effect of time and a significant effect of condition, but no significant interaction effect. We can interpret these results with the help of Fig. 2.[13] According to the analysis, psychomotor vigilance was reduced significantly over days of sleep deprivation (as expected); and the placebo condition performed consistently worse than the caffeine condition. This latter finding included Day 0, before the beginning of pill administration. Any additional difference between conditions due to the action of caffeine (during Days 1–3, but not Day 0) should have led to an interaction effect in this analysis, but no significant interaction was found. There is relatively little statistical

TABLE II
RESULTS FROM MIXED-DESIGN REPEATED-MEASURES ANOVA[a]

Effect	SS	df	MS	F	p
Condition (C)	1082.8	1	1082.8	8.03	0.009
Error (E)	3235.1	24	134.8		
Time (A)	3745.4	3	1248.5	39.88	<0.001
Interaction (I)	83.4	3	27.8	0.89	0.452
Error (R)	2253.9	72	31.3		

[a] Results are shown from repeated-measures ANOVA of psychomotor vigilance performance data over 4 days of sleep deprivation in two different conditions. The sums of squares (SS), the degrees of freedom (df), and the means of squares (MS) for the different variance components are displayed. In addition, the F statistics and p values for the effect of condition, effect of time, and interaction of condition by time are given.

[13] The use of figures or tables of the data is crucial for the interpretation of time effects and interactions, because the results of repeated-measures ANOVA do not reveal the direction of changes in the means over time or between conditions. Moreover, repeated-measures ANOVA is insensitive to the order of the time points; if the data points of an upward effect are rearranged (in the same manner for each subject) to form a downward effect, the ANOVA results remain the same. Repeated-measures ANOVA also does not take into account the intervals between the time points; for typical applications, the interval between all adjacent time points would be considered the same.

power in interaction effects,[14] however, making it difficult to conclude with any degree of certainty that caffeine was ineffective in this experiment.[15]

If the substantial interindividual variability in psychomotor vigilance performance impairment during total sleep deprivation (Fig. 1) is systematic, as has been previously reported,[16] then there is reason to believe that the repeated-measures ANOVA results are inaccurate. Repeated-measures ANOVA does not distinguish variance due to systematic interindividual differences from random error variance,[11] lumping these together as a single source of variance. As a consequence, the result for the effect of condition in the current analysis is unreliable, as is evident from the expression for SS(E) in Eq. (12). Depending on the particular design, interindividual variability may lead to overestimation or underestimation of statistical significance.[17] In the present mixed design, the effect of condition is underestimated due to systematic interindividual differences. Therefore, other analyses are warranted to further investigate this data set.

Mixed-Model Regression Analysis

Regression analysis is essentially equivalent to ANOVA; while ANOVA focuses on the variance in the data to assess differences between the means of subsets of the data, however, regression analysis focuses on assessing the parameters of a model (i.e., mathematical function) posited to describe the data set. Depending on the criteria used to determine the optimal parameter values, regression analysis typically involves minimization of the error variance[18] (which, as we shall see, presents a methodological connection between regression analysis and ANOVA[19]).

[14] B. J. Winer, "Statistical Principles in Experimental Design," 2nd Ed., McGraw-Hill, New York, NY, 1971.

[15] By expressing the data on Days 1–3 as relative to those on Day 0, the difference between the two conditions prior to pill administration could be eliminated. Any consistent difference due to caffeine (i.e., present throughout Days 1–3) should then result in a main effect of condition, which is statistically more powerful than an interaction effect. On the other hand, by expressing the data as relative to those on Day 0, the noise in the data for Day 0 would be propagated to the data for Days 1–3. For the present data, it effectively makes no difference in study outcomes (effect of time: $F[2, 48] = 15.64, p < 0.001$; effect of condition: $F[1, 24] = 0.35, p = 0.559$; interaction effect: $F[2, 48] = 1.04, p = 0.362$).

[16] H. P. A. Van Dongen, N. L. Rogers, and D. F. Dinges, *Sleep Biol. Rhythms* **1,** 5 (2003).

[17] H. A. Feldman, *J. Appl. Physiol.* **64,** 1721 (1988).

[18] In standard regression analysis, as in ANOVA, minimization of the error variance can be achieved by means of the least-squares method (i.e., minimization of the squares of the deviations between the model predictions and the data points). In mixed-model regression analysis, (approximate) minimization of the error variance is rather a by-product of the maximum likelihood approach used to estimate the model parameters.

Regression analysis provides a richer framework than ANOVA, in that a wider variety of models for the data can be evaluated.[20] We focus here on mixed-model (or mixed-effects) regression analysis,[21] which means that the model posited to describe the data contains both fixed effects and random effects. Fixed effects are those aspects of the model that (are assumed to) describe systematic features in the data. Fixed effects are used to determine expected or mean values for the subject population (as such, they can be compared to the regression coefficients in a standard regression analysis on pooled data, or to the effects of condition, time and interaction in repeated-measures ANOVA). Random effects are those aspects of the model that are allowed to vary among subjects (i.e., parameters that take different values depending on which individual subject the data are from). Random effects are variance components that describe the variability (e.g., biological variability) in the observations around the expected values as predicted by the fixed effects. In this chapter, random effects will be indicated by Greek letters, to distinguish them from fixed effects.

To estimate model parameters in a standard regression analysis, the least-squares method, which involves minimization of the error variance, can be used. For mixed-model regression analysis, which requires more complicated computations to be made, an alternative method for parameter estimation is used: maximum likelihood estimation.[18] The basic idea underlying maximum likelihood estimation is that the data reflect the most probable outcome under the conditions specified in the model. Thus, given certain assumptions about the statistical distribution(s) of the stochastic process(es) involved in the mechanisms that generated the data (such as random between-subjects variability or noise), maximum likelihood estimation involves finding the model parameter values that maximize the likelihood of observing the data at hand.

To put this in mathematical equations, let us consider a mixed-effects model of the form

$$y_{jit} = f_{jt} + \eta_{ji} + \varepsilon_{jit} \tag{19}$$

[19] Note that regression analysis is not contingent upon having a complete data set. In traditional repeated-measures ANOVA, however, a missing data point eliminates the entire subject from the data set.

[20] In contrast with repeated-measures ANOVA, regression analyses usually involve models that are sensitive to the order of the time points, and take into account the intervals between the time points.

[21] There are other regression techniques that could have utility with regard to the present data, such as analysis of covariance (ANCOVA). However, ANCOVA is a fixed-effects method, and it is more restricted than mixed-effects regression analysis.[17]

where f_{jt} is a function of t representing the fixed effects posited to describe the data at the group level (i.e., for conditions j overall). The term η_{ji} represents the random effect—more precisely, the η_{ji} are the subject-specific instances of the random effect, for subjects i in conditions j (which are lumped together here), that are usually assumed to arise from some family of parametric probability distributions such as a normal distribution with zero mean and variance ω^2 over subjects (where ω^2 is to be estimated). For the present purpose, the random effect is assumed to be additive to the fixed effects f_{jt}. The ε_{jit} represent independent noise assumed to have a normal distribution with zero mean and variance σ^2 over subjects (where σ^2 is to be estimated) for all times t. Further, the distributions of η and ε are assumed to be independent (i.e., zero covariance). Let us temporarily assume there is only one group j, so that Eq. (19) can be simplified to

$$y_{it} = f_t + \eta_i + \varepsilon_{it} \tag{20}$$

For each subject i, the likelihood l_i of observing the subject's time series data y_{it} is given by

$$l_i = \Pi_t \, c \, N[f_t + \eta_i; \sigma^2](y_{it}) \tag{21}$$

where Π_t denotes multiplication over t (equivalent to Σ_t for summation), and c is a (here irrelevant) normalization factor. $N[f_t + \eta_i; \sigma^2]$ is the density function for a normal distribution with mean $f_t + \eta_i$ and variance σ^2; in Eq. (21) this function is evaluated at the data points y_{it}. Assuming that the η_i are taken from a normal distribution $N[0; \omega^2](\eta_i)$ with zero mean and variance ω^2 over subjects, we can define the marginal likelihood L_i to integrate η_i out:

$$L_i = C \int_{\eta_i = -\infty}^{\infty} l_i \cdot N[0; \omega^2](\eta_i) d\eta_i \tag{22}$$

where C is a normalization factor. The likelihood L of observing the entire data set is then given by

$$L = \Pi_i L_i \tag{23}$$

The likelihood L is a function of the parameters that contributed to its derivation: the (currently unspecified) parameters constituting the fixed effects f_t, the (between-subjects) variance ω^2 of the random effect, and the (within-subjects) error variance σ^2. By maximizing the value of L, maximum likelihood estimates of these parameters are obtained.[22] In practice, computer software is used to estimate the parameters, as numerical approximation is needed to find the parameter values that maximize L.

Linear Mixed-Model Regression Analysis Applied to the Data:
Mixed-Model ANOVA

To illustrate the use of mixed-model regression analysis in practice, we
first replicate the repeated-measures ANOVA performed previously, using
the data presented in Fig. 2. The only essential difference with the
repeated-measures ANOVA is, of course, that we take systematic interindividual differences into account. Because of the equivalence with
repeated-measures ANOVA, this application of mixed-model regression
analysis is also known as mixed-model ANOVA. The procedure involves
explicit estimation of the model parameters, however, by means of the
maximum likelihood method outlined previously.

Let us define day indicator variables d_{tu}, which equal 1 if $t = u$ and 0
otherwise, and condition indicator variables c_{jk}, which equal 1 if $j = k$
and 0 otherwise ($j = 1$ corresponds to the caffeine condition). We consider
the following linear mixed-effects model for the data:

$$y_{jit} = I_{ji} + d_{t1}Y_1 + d_{t2}Y_2 + d_{t3}Y_3 + c_{j1}(d_{t0}Z_0 + d_{t1}Z_1 + d_{t2}Z_2 + d_{t3}Z_3) + \varepsilon_{jit}$$

(24)

where parameter I_{ji} is the intercept (which represents the mean for Day 0
in the placebo condition), involving a random effect such that

$$I_{ji} = I_0 + \eta_{ji}$$

(25)

The η_{ji} constitute the random effect for subjects i in conditions j (which are
lumped together), assumed to arise from a normal distribution with zero
mean and variance ω^2 over subjects (where ω^2 is not known in advance).
Parameters Y_1, Y_2, and Y_3 represent the means for Days 1, 2, and 3 in the placebo condition, respectively, expressed as differences from the intercept;
and parameters Z_0 through Z_3 represent the means for Days 0 through 3 in
the caffeine condition, respectively, expressed as differences from their
counterparts in the placebo condition. As an example, the data point for
subject 5 in the caffeine condition (i.e., condition 1) on Day 2 is modeled as

---

[22] Maximum likelihood (ML) parameter estimates tend to be (slightly) biased.[17] An improved
methodology called restricted maximum likelihood (REML) is available for linear mixed-
model regression analysis (but not for nonlinear mixed-model regression analysis). REML
provides unbiased parameter estimates that are preferable to those resulting from
conventional ML in virtually all cases (except when comparing models with different
fixed-effects structures on the basis of the likelihood ratio χ^2 statistic[25]). In this chapter, we
report REML parameter estimates for all linear mixed-effect regression analyses, and ML
parameter estimates for all nonlinear mixed-model regression analyses (for which REML is
not available). For more on this topic, see P. J. Diggle, K.-Y. Kiang, and S. L. Zeger,
"Analysis of Longitudinal Data." Clarendon Press, Oxford, 1996.

$$y_{152} = I_0 + \eta_{15} + Y_2 + Z_2 + \varepsilon_{152} \tag{26}$$

as all other terms in Eq. (24) cancel out.

We assume that we have the psychomotor vigilance performance data for Days 0–3 of the sleep deprivation experiment stored in a spreadsheet, organized in columns of $(n_1 + n_2)\, T = 26 \times 4 = 104$ rows each: a column listing the different subjects (named "id"), a column equivalent to the indicator variable c_{j1} indicating in which condition each subject was (named "cond"), four columns equivalent to the indicator variables d_{tu} (named "day0" through "day3") marking the day in the experiment (such that column "day$u$" equals 1 if the data in that row are from day u and 0 otherwise), and a column for the data of each subject on each day (named "y"). For (nonessential) technical reasons, the spreadsheet must be ordered by subject (i.e., all the data for a single subject must appear consecutively in the spreadsheet).[23] To perform the calculations for this linear mixed-model regression analysis, we use the following PROC MIXED command in the computer software SAS[24]:

```
proc mixed;
    class id;
    model y = day1 day2 day3 day0*cond day1*cond day2*cond day3*cond
        /solution;
    random intercept /solution subject=id;
run;
```

The second line of this command specifies a categorical variable by which the data are classified—since repeated measures were obtained for each subject, we classify the data by subject (column "id"). The third line codes for the fixed effects in the model of Eq. (24). In the PROC MIXED command it is unnecessary to explicitly specify the parameters of the model; each term on the second line automatically has a parameter associated with it (e.g., "day1*cond" refers to Z_1, which is automatically multiplied by the value of column "day1" times the value of column "cond," i.e., by $c_{j1}\, d_{t1}$). Furthermore, unless otherwise specified, an intercept is automatically assumed to be included in the model. The "solution" option requests that the parameter estimates be reported. The fourth line puts a

---

[23] Mixed-model regression analyses are robust against random deviations from a balanced design (i.e., different numbers of observations among subjects) and randomly occurring missing values. However, the PROC MIXED command in SAS expects missing values to be indicated in the data spreadsheet by means of periods, so that the location of the missing values relative to the available observations is clear. This can be circumvented by using the "repeated" statement in the PROC MIXED command.[25]

[24] SAS System for Windows, release 8.02. SAS Institute Inc., Cary, NC, 2001.

random effect on the intercept, with the random elements being the different subjects (column "id") as specified in the "subject=" option. The "solution" option requests that the empirical best linear unbiased predictors (EBLUPs) of the random effect for the individual subjects (i.e., the estimates for the η_{ji}) be reported. The "run" statement in the last line requests execution of the analysis.

The previous command represents a complete analysis, but in order to get results equivalent to those obtained with repeated-measures ANOVA, a few statements must be added to the code:

```
proc mixed;
  class id;
  model y = day1 day2 day3 day0*cond day1*cond day2*cond day3*cond
    /solution;
  random intercept /solution subject=id;
  estimate "effect of condition" day0*cond 0.25 day1*cond 0.25
                               day2*cond 0.25 day3*cond 0.25;
  contrast "effect of time" day1 1 day1*cond 0.5 day0*cond -0.5,
                            day2 1 day2*cond 0.5 day0*cond -0.5,
                            day3 1 day3*cond 0.5 day0*cond -0.5;
  contrast "interaction effect" day1*cond 1 day0*cond -1,
                                day2*cond 1 day0*cond -1,
                                day3*cond 1 day0*cond -1;
run;
```

The contrast statements, appropriately labeled "effect of time" and "interaction effect," are equivalent to those effects in repeated-measures ANOVA. The contrast for the effect of time again constitutes a test that the means over all subjects (lumping the two conditions) are equal for all time points (i.e., for Day 0 versus Day 1, for Day 0 versus Day 2, and for Day 0 versus Day 3). As the number of subjects in each condition is the same, this translates into the following three-fold null hypothesis:

$$H_0 : \begin{cases} I_0 + Z_0/2 = I_0 + Y_1 + Z_1/2 \\ I_0 + Z_0/2 = I_0 + Y_2 + Z_2/2 \\ I_0 + Z_0/2 = I_0 + Y_3 + Z_3/2 \end{cases} \tag{27}$$

which can be simplified as follows:

$$H_0 : \begin{cases} Y_1 + Z_1/2 - Z_0/2 = 0 \\ Y_2 + Z_2/2 - Z_0/2 = 0 \\ Y_3 + Z_3/2 - Z_0/2 = 0 \end{cases} \tag{28}$$

The latter form of the null hypothesis is coded in the contrast statement for the effect of time shown previously. For details about the formulation of the "contrast" statement in the PROC MIXED command, see the online SAS user's guide.[25]

As with its equivalent in repeated-measures ANOVA, the contrast for the interaction effect is essentially a test that the difference between the condition means is the same for all time points (i.e., for Day 0 versus Day 1, for Day 0 versus Day 2, and for Day 0 versus Day 3). This translates into the following null hypothesis:

$$H_0 : \begin{cases} Z_0 = Z_1 \\ Z_0 = Z_2 \\ Z_0 = Z_3 \end{cases} \tag{29}$$

which can be reformulated as follows:

$$H_0 : \begin{cases} Z_1 - Z_0 = 0 \\ Z_2 - Z_0 = 0 \\ Z_3 - Z_0 = 0 \end{cases} \tag{30}$$

The latter form of the null hypothesis is coded in the contrast statement for the interaction effect shown previously.

In repeated-measures ANOVA, the effect of condition represents a test of whether the grand means are the same for the different conditions. In the context of Eq. (24), this corresponds to a test of the following null hypothesis:

$$H_0 : \begin{aligned} &[I_0 + (I_0 + Y_1) + (I_0 + Y_2) + (I_0 + Y_3)]/4 = \\ &[(I_0 + Z_0) + (I_0 + Y_1 + Z_1) + (I_0 + Y_2 + Z_2) + (I_0 + Y_3 + Z_3)]/4 \end{aligned} \tag{31}$$

which can be reduced to

$$H_0 : [Z_0 + Z_1 + Z_2 + Z_3]/4 = 0 \tag{32}$$

This leads directly to a simple test for the effect of condition, namely through evaluation of the estimated value for $[Z_0 + Z_1 + Z_2 + Z_3]/4$, the actual difference between the grand means for the two conditions. This test for the effect of condition is different, and more powerful, than the one available in repeated-measures ANOVA. Moreover, it is robust against systematic interindividual differences, as these are absorbed by the random effect for the intercept. In the PROC MIXED command shown previously, the test is implemented by means of the "estimate" statement,[25] labeled "effect of condition" here. It yields the estimated value of the expression

[25] SAS OnlineDoc SAS/STAT User's Guide, version 8. SAS Institute Inc., Cary, NC, 2001.

at hand and automatically performs a t test against zero, with the following degrees of freedom[26]:

$$df_c = \Sigma_j(n_j - 1)(T - r) \tag{33}$$

where r is the number of random effects (which equals 1 in the present model).

Table III shows the results of the linear mixed-model regression analysis. The effect of time and the interaction effect are identical to those found for repeated-measures ANOVA (cf. Table II). The effect of condition is different, however, as was to be expected. Provided that the assumption of a normal distribution for the random effect is correct, in mixed-model regression analysis the effect of condition is not influenced by systematic interindividual differences. Still, even with this improved test for the effect of condition, the overall results are the same as for repeated-measures ANOVA: a significant effect of time and a significant effect of condition, but no significant interaction effect. Thus, psychomotor vigilance was reduced significantly over days of sleep deprivation; and the placebo condition performed consistently worse than the caffeine condition, as was already clear on Day 0 (before pill administration began).

An Alternative Linear Mixed-Effects Model

As mentioned earlier, any effect of caffeine (during Days 1–3) on psychomotor vigilance should have led to a significant interaction effect. However, since there is relatively little statistical power in interaction effects,[14]

TABLE III
RESULTS FROM MIXED-MODEL ANOVA[a]

Effect	Test	Statistic	df	p
Condition	t	2.83	1	0.006
Time	F	39.88	3, 72	<0.001
Interaction	F	0.89	3, 72	0.452

[a] Results are shown from mixed-model regression analysis, mimicking repeated-measures ANOVA, on psychomotor vigilance performance data over 4 days of sleep deprivation in two different conditions. The type of test (F or t test), the value of the test statistic, the degrees of freedom, and the p value are given for the effect of condition, effect of time, and interaction of condition by time. These effects are defined equivalently to those for repeated-measures ANOVA (cf. Table II).

[26] The resulting t statistic, when squared, yields the F statistic for an equivalent F test with 1, df_c degrees of freedom.

the results from this mixed-model ANOVA (Table III) could not be much more helpful than the results from the repeated-measures ANOVA (Table II) given that the interaction effect is identical for the two approaches. Yet, the mixed-effects model in Eq. (24) also yields a priori day-by-day comparisons between the conditions (i.e., the parameters Z_0 through Z_3). These same comparisons would require post-hoc tests in the repeated-measures ANOVA approach,[27] which again would yield results confounded by systematic interindividual variability in the data.

The "solution" option for the "model" statement in the PROC MIXED command shown previously produces the parameter estimates for the fixed effects in the model, and automatically performs a t test against zero (with df_c degrees of freedom) for each. The fixed effects "day0*cond" through "day3*cond" are of interest, since they correspond to the parameters Z_0 through Z_3. The results are shown in Table IV, revealing a significant difference between the caffeine and placebo conditions on Day 1 only. This finding would suggest that the effect of sustained low-dose caffeine was limited to the first day of administration. However, we are still faced with the systematic (albeit nonsignificant) difference between conditions on Day 0 (i.e., prior to pill administration; see Fig. 2). This makes it difficult to tell to what extent the difference between conditions on Day 1 might be nonspecific to caffeine.

The substantial flexibility we have in formulating a model for mixed-model regression analysis is helpful to address this problem. As assignment

TABLE IV
FIXED EFFECTS IN MIXED-MODEL ANOVA[a]

Day	Mixed-model ANOVA		Repeated-measures ANOVA	
	$t[72]$	p	$t[24]$	p
0	1.87	0.066	2.79	0.010
1	3.20	0.002	2.95	0.007
2	1.96	0.053	2.03	0.054
3	1.67	0.100	1.39	0.177

[a] The table shows statistical tests of the differences in psychomotor vigilance performance between conditions for each of the 4 days of total sleep deprivation. The t statistics (with 72 degrees of freedom) and p values for these differences as resulting from mixed-model ANOVA are given. For comparison, the t statistics (with 24 degrees of freedom) and p values for the equivalent (post-hoc) tests in repeated-measures ANOVA are also shown.[27] These latter results are confounded by the systematic interindividual variability in the data.

[27] In SPSS, these specific tests can be obtained with the "parameter estimates" option for the repeated-measures ANOVA procedure.

to condition in this double-blind study was random, the difference between conditions before pill administration should be random and unrelated to condition. We may therefore consider a slightly modified (and more parsimonious) mixed-effects model for the data:

$$y_{jit} = I_{ji} + d_{t1}Y_1 + d_{t2}Y_2 + d_{t3}Y_3 + c_{j1}(d_{t1}Z_1 + d_{t2}Z_2 + d_{t3}Z_3) + \varepsilon_{jit} \quad (34)$$

which is identical to the model in Eq. (24) except that the Z_0 term for the difference between conditions on Day 0 is left out. The estimated means for the two conditions on Day 0 in this model are a function of the EBLUPs for the random effect on the intercept in Eq. (25), as follows:

$$m_{j0} = I_0 + \Sigma_i \, \eta_{ji}/n_j \quad (35)$$

Since the mean of all the EBLUPs should be (almost) identical to zero (recall that the η_{ji} are assumed to arise from a normal distribution with zero mean), it follows that $(m_{10} + m_{20})/2 = I_0$.

To perform the calculations of the mixed-model regression analysis for Eq. (34), we use the following PROC MIXED command in the computer software SAS:

```
proc mixed;
   class id;
   model y = day1 day2 day3 day1*cond day2*cond day3*cond /solution;
   random intercept /solution subject=id;
run;
```

The EBLUPs are generated by the "solution" option to the "random" statement. They are shown in Table V; substantial interindividual differences are apparent. Using Eq. (35) and the parameter estimates for Eq. (34), we derive the estimated means for each day in each of the two conditions to get a sense of how well the model fits the data. The results are shown in Fig. 4. As expected, this model fits the means as accurately as the model of Eq. (24) or the repeated-measures ANOVA approach (which both provided a perfect fit to the means; cf. Fig. 2), except on Day 0. This suggests that the distribution of the random elements (i.e., the subjects' performance on Day 0) is not precisely normal.[28]

We make use again of the a priori day-by-day comparisons between the conditions (i.e., fixed-effect parameters Z_1 through Z_3) in the model of Eq. (34) to confirm the earlier tentative finding that the effect of sustained low-dose caffeine was limited to the first day of pill administration. The estimated values for Z_1 through Z_3 and the t tests for whether they differ significantly from zero are shown in Table VI. These results confirm the significant difference between the caffeine and placebo conditions on Day 1 only,

TABLE V

EBLUPs Resulting from Mixed-Model ANOVA[a]

id	cond	EBLUP
1	1	1.7
2	1	0.0
3	1	−0.8
4	1	−0.4
5	1	−0.4
6	1	0.4
7	1	−2.3
8	1	−1.1
9	1	−1.0
10	1	−2.5
11	1	−4.1
12	1	−3.0
13	1	−3.4
14	2	1.5
15	2	5.4
16	2	−2.6
17	2	10.4
18	2	7.4
19	2	−11.3
20	2	−7.6
21	2	−3.0
22	2	1.5
23	2	5.4
24	2	2.5
25	2	0.2
26	2	6.9

[a] Empirical best linear unbiased predictors (EBLUPs) for the η_{ji} in the mixed-effects regression model of Eq. (34), representing subject-specific deviations in the intercept relative to the overall intercept I_0. The EBLUPs are given in number of psychomotor vigilance performance lapses per 10-min test bout, for each of the subjects (column "id") in the caffeine condition (column "cond" value 1) and the placebo condition (column "cond" value 2).

[28] The estimate for I_0 is 4.9 ± 1.5 (mean ± standard error). Inspection of Table V reveals that the estimated subject-specific performance $I_0 + \eta_{ji}$ on Day 0 is negative (i.e., less than zero lapses per 10-min test bout) for some subjects ($i = 19, 20$). This anomaly again suggests that the distribution of the random elements is not precisely normal. Specification of other types of distributions for the random effect(s), such as the lognormal distribution that always yields positive values, is possible in SAS.[25] However, the results of mixed-effects modeling are not critically dependent on the assumptions about the distribution of the random effect(s), especially if many repeated measures are available. See E. Olofsen, D. F. Dinges, and H. P. A. Van Dongen, *Aviat. Space Environ. Med.* (Suppl.), in press.

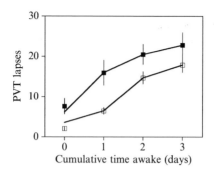

FIG. 4. Estimates for the daily means of lapses on the psychomotor vigilance task (PVT) for the caffeine condition (lower curve) and the placebo condition (upper curve), derived from the model in Eq. (34). The boxes represent condition means (with standard errors of the mean) for the actual data (see Fig. 2).

TABLE VI
DAY-BY-DAY COMPARISONS IN LINEAR MIXED-MODEL REGRESSION ANALYSIS[a]

Day	Z	t[72]	p
1	6.9	2.58	0.012
2	3.2	1.21	0.231
3	2.4	0.88	0.382

[a] Differences in psychomotor vigilance performance between conditions are shown for each of the 3 days of total sleep deprivation following the beginning of pill administration (i.e., Days 1–3), as assessed with the adjusted mixed-effects regression model of Eq. (34). The differences in the number of lapses Z and the corresponding t statistics (with 72 degrees of freedom) and p values are given.

independently of the systematic interindividual differences in the data (i.e., independently of the random effect). Thus, we have gained more definitive evidence that the effect of sustained low-dose caffeine was limited to the first day of intake, despite the fact that plasma caffeine concentrations were high throughout the 66 h of pill administration (Fig. 1).

Nonlinear Mixed-Model Regression Analysis

We now shift our attention from linear mixed-model regression analysis to nonlinear mixed-model regression analysis. Although linear and non-linear mixed-effects models are formulated quite differently in most published literature and computer software (e.g., see the online SAS user's guide[25]), they are actually intimately related, linear mixed-effects modeling being a special case of nonlinear mixed-effects modeling (for this reason, we have standardized the notation throughout this chapter). Nonlinear mixed-model regression is frequently needed to analyze hypothesis-driven models (i.e., models that go beyond describing the data in terms of unspecified changes over time and/or differences among conditions as in ANOVA), as such models tend to include nonlinear combinations of fixed and/or random effects.[29] The extensive numerical calculations required for nonlinear mixed-model regression analysis have become feasible in the last 5 years due to the increasing computational power of standard computer hardware. We can take advantage of this development for the further analysis of our study data.

Considering the evidence we gathered thus far that the attenuation of performance impairment by caffeine dissipated over days of sleep deprivation, we wonder about the precise duration of the efficacy of sustained low-dose caffeine in this experiment. We therefore consider the following mixed-effects model for the study data as a function of days t[30]:

$$y_{ji}(t) = I_{ji} + b_0 t^s - [(1 - d_{t0}) c_{j1} z(t)] + \varepsilon_{ji}(t) \tag{36}$$

where model parameters I and ε and indicator variables c and d are defined as in the previous sections. The function $b_0 t^s$ has previously been shown to describe the data in the placebo condition[31]; it involves a curvature parameter $0 < s < 1$ and a scale factor b_0. The term between square brackets represents the hypothesized temporal change in caffeine's efficacy—for the caffeine condition only and on Days 1–3 only (i.e., during caffeine pill administration). We hypothesize that beginning with the first pill (i.e., on Day 1), the efficacy $z(t)$ of hourly administration of caffeine diminishes exponentially over days t:

$$z(t) = ae^{-(t-1)/T_0} \tag{37}$$

where a is a scale factor and T_0 is a time constant for the decline of caffeine's efficacy. The model of Eq. (36), though linear in the random effect

---

[29] Hybrid models, containing mixed-model ANOVA elements as well as hypothesis-driven components, can be readily constructed from the formulas in this chapter, and typically require nonlinear mixed-model regression analysis as well.

[30] It is generally advisable to centralize the independent (and dependent) variables of a regression analysis (i.e., adding constants to each so that the ranges of values they take center on zero). This practice tends to reduce the covariance among model parameters (especially in linear regression models) and promotes the reliability of model convergence. In Eq. (36), however, centralization is problematic for the independent variable t, since the term $b_0 t^s$ would be undefined for $t < 0$.

[31] H. P. A. Van Dongen, G. Maislin, J. M. Mullington, and D. F. Dinges, *Sleep* **26,** 117 (2003).

included in the intercept I_{ji} of Eq. (25), cannot be cast in the form of a linear mixed-model regression model, and must be subjected to nonlinear mixed-model regression analysis.

Let us assume that we have the psychomotor vigilance performance data for Days 0–3 of the sleep deprivation experiment stored in a spreadsheet similar to that described in the previous sections, with a column listing the different subjects (named "id"), a column indicating the days (named "t"), a column equivalent to the indicator factor $[(1 - d_{t0}) c_{j1}]$ (named "caff"), and a column for the data of each subject on each day (named "y").[32] For technical reasons, the spreadsheet must again be ordered by subject. To perform the calculations for the nonlinear mixed-model regression analysis, we use the following PROC NLMIXED command in the computer software SAS:

```
proc nlmixed;
  parms
     i0 = 5.0,
     s2i = 25.0,
     b0 = 10.0,
     a = 20.0,
     s = 0.5,
     t0 = 1.0,
     s2e = 35.0;
     z = a*caff*exp (-(t - 1)/t0);
  if t = 0 then v = 0;
     else v = t**s;
  model y ~ normal (i0 + vari + b0*v - z, s2e);
  random vari ~ normal(0,s2i) subject=id out=ebes;
run;
```

The "parms" part of this command introduces the seven parameters of the model, and their initial values[33] (which the computer software requires to begin the calculations). The parameters are "i0" for the intercept I_0, "s2i" for the variance ω^2 of the random effect η for the intercept as in Eq. (25), "b0" and "a" for scale factors b_0 and a in Eqs. (36) and (37), respectively, "s" for the curvature parameter s, "t0" for the time-constant T_0, and "s2e"

---

[32] For model convergence and reliability of analysis outcomes in PROC NLMIXED, it is desirable that the dependent variable y and the independent variables of the model (in this case only t) have comparable ranges (i.e., same order of magnitude). Linear transformations should be used as necessary to accomplish this.

[33] Depending on the complexity of the model, the choice for the initial values can be critical for model convergence and for the success of the analysis. Proper initial values can often be derived from a two-stage analysis of the same data.[2,17,41]

for the variance σ^2 of the error term ε. These are all the parameters explicitly and implicitly contained in the model of Eq. (36). The "z= ... " line in the command computes the function $z(t)$ of Eq. (37). The "if t=0 ... " part of the command is necessary because SAS does not adopt the convention that $t^s = 0$ for $t = 0$. Thus, we introduce a substitute variable "v" defined by $v = t^s$ for $t > 0$ and $v = 0$ for $t = 0$; v replaces t^s in Eq. (36) without changing the model.

The "model" statement defines the representation for the data y as normally distributed random fluctuations with variance σ^2 (i.e., "s2e") around the model of Eq. (36). That model is described by the following code in the "model" statement:

```
i0 + vari + b0*v - z
```

which follows directly from Eqs. (36) and (37), except for the term "vari" that represents the random effect for the intercept. This random effect is defined in the "random" statement as a normal distribution with mean zero and variance ω^2 (i.e., "s2i"); the declaration "subject=id" specifies that the random effect pertains to variability among subjects. Finally, the "out=" option stores the empirical Bayes estimates (EBEs) for the random effect η_{ji} (the equivalent of the EBLUPs in linear mixed-model regression analysis) in a spreadsheet called "ebes."

Table VII shows the parameter estimates resulting from the nonlinear mixed-model regression analysis. Of primary interest is the estimate for T_0 (i.e., 1.2336 before rounding), from which we can derive the half-life $T_{0.5}$ of the efficacy of sustained low-dose caffeine using the following expression:

$$e^{-T_{0.5}/T_0} = 0.5 \qquad (38)$$

It follows that $T_{0.5} = 0.86$ days. Thus, it appears that sustained low-dose caffeine lost half of its efficacy in less than a day, which is consistent with what we derived using linear mixed-model regression analysis (see previously). This finding could reflect a rapid build-up of tolerance to caffeine. Alternatively, the build-up of sleepiness during the extended sleep deprivation could have simply overwhelmed the stimulating effect of caffeine after about a day.

The relatively large standard error in Table VII for the T_0 estimate would seem to indicate that the effect of sustained low-dose caffeine in this experiment may not be very robust. In fact, the estimate for T_0 is not significantly different from zero ($t[25] = 1.08, p = 0.29$). This warrants investigation of whether removal from the model of this parameter, and thereby the entire function $z(t)$, would constitute a significant deterioration in how well the model describes the data. This can be assessed with the likelihood ratio test, which involves calculation of -2 times the natural logarithm of the likelihood (i.e., the value of $-2 \log L$) for the full model (with all

TABLE VII
PARAMETER ESTIMATES FOR NONLINEAR MIXED-EFFECTS REGRESSION MODEL[a]

Parm	Name	Est	SE
I_0	i0	4.8	1.5
ω^2	s2i	27.0	10.1
b_0	b0	10.1	1.9
a	a	7.0	2.6
s	s	0.44	0.15
T_0	t0	1.2	1.1
σ^2	s2e	29.5	4.8

[a] Results are shown from nonlinear mixed-model regression analysis of psychomotor vigilance performance data over 4 days of sleep deprivation, using the hypothesis-driven model in Eq. (36) to assess the duration of the efficacy of sustained low-dose caffeine (relative to placebo). As computed using the PROC NLMIXED command in SAS, the table shows the parameters (Parm), their names in the command (Name), their estimates (Est), and their standard errors (SE).

parameters included) and for the reduced model (with the parameters that might be unnecessary being removed). By subtracting the $-2 \log L$ value for the full model from the corresponding value for the reduced model, the likelihood ratio is computed. This statistic approximately has a χ^2 distribution, and the difference in the number of free parameters between the full and reduced models determines the degrees of freedom for that χ^2 distribution (see the online SAS user's guide[25]). Using the likelihood ratio test, we find that inclusion of parameter T_0 (and thereby also parameter a) results in a significant improvement over the model without T_0 ($\chi^2[2] = 7.50, p = 0.024$).

It is useful also to graphically check how well the nonlinear mixed-effects model of Eq. (36) fits the data. Using Eq. (35), we first estimate the means for the two conditions on Day 0 in this model, which are a function of the EBEs for the random effect on the intercept in Eq. (36). The EBEs are shown in Table VIII; they are similar to those found for the linear mixed-effects regression model of Eq. (34) (cf. Table V). Figure 5 shows the nonlinear mixed-effects model overlaid on the group mean data. It appears that the model in Eq. (36) fits the means well, except on Day 0. As in the model of Eq. (34), this suggests that the distribution of the random effect for the intercept among subjects is not precisely normal. However, Fig. 5 is reassuring with regard to the approximate validity of the hypothesized dissipation profile in Eq. (37), at least as it pertains to the group means.

An explanation for the relatively large standard error for the time constant T_0 might be that an additional random effect is needed in the model. Previous analyses have revealed large variability in the scale factor b_0 for the placebo condition.[31] Therefore, we add a random effect to the model in Eq. (36) as follows:

TABLE VIII
EBEs Resulting from Nonlinear Mixed-Model
Regression Analysis[a]

id	cond	EBE
1	1	1.6
2	1	−0.1
3	1	−0.9
4	1	−0.5
5	1	−0.5
6	1	0.3
7	1	−2.5
8	1	−1.2
9	1	−1.1
10	1	−2.6
11	1	−4.2
12	1	−3.1
13	1	−3.5
14	2	1.6
15	2	5.6
16	2	−2.5
17	2	10.6
18	2	7.5
19	2	−11.3
20	2	−7.5
21	2	−2.9
22	2	1.6
23	2	5.6
24	2	2.6
25	2	0.3
26	2	7.1

[a] Empirical Bayes estimates (EBEs) for the η_{ji} in the nonlinear mixed-effects regression model of Eq. (36), representing subject-specific deviations in the intercept relative to the overall intercept I_0. The EBEs are given in number of psychomotor vigilance performance lapses per 10-min test bout, for each of the subjects (column "id") in the caffeine condition (column "cond" value 1) and placebo condition (column "cond" value 2).

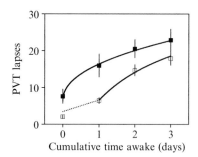

FIG. 5. Hypothesis-driven model for the performance-impairing effect of total sleep deprivation and the transient mitigating effect of caffeine, as measured by the daily means of lapses on the psychomotor vigilance task (PVT). The model for the placebo condition is shown by the upper curve, and the model for the caffeine condition is shown by the lower curve. The dotted part of the lower curve connects the period prior to caffeine administration with the period during which caffeine was administered; the boundary between these two periods involves a discontinuity in the model. The boxes represent condition means (with standard errors of the mean) for the actual data (Fig. 2).

$$y_{ji}(t) = I_{ji} + b_{ji}\,t^s - [(1 - d_{t0})c_{j1}z(t)] + \varepsilon_{ji}(t) \qquad (39)$$

where

$$b_{ji} = b_0 + \beta_{ji} \qquad (40)$$

The β_{ji} are assumed to arise from a normal distribution with zero mean and variance λ^2 over subjects (with λ^2 not known in advance). The model now has two random effects (one for the intercept I and one for the scale factor b); we assume that the covariance between these two random effects is zero.[34]

Even though the PROC NLMIXED command in SAS[24] can deal with two (but not more than two) random effects, it is now more convenient to use the specialized computer software NONMEM[35] to perform the calculations for the nonlinear mixed-model regression analysis.[36] Using the same data in the same spreadsheet (named "CAFF.DAT") as for PROC NLMIXED in SAS, ordered by subject, we apply the following NONMEM macro:

---

[34] For small subject populations, the covariance between random effects is usually not well estimable; inappropriately setting it to zero is probably no more problematic than estimating it poorly.

[35] NONMEM version V level 1.1. GloboMax LLC, Hanover, MD, 1998.

[36] We have found that in NONMEM[35] the numerical computations for models with two random effects are more likely to converge, over a wider range of initial values for the parameters.

```
$PROBLEM CAFFEINE
$DATA CAFF.DAT
$INPUT ID T CAFF DV
$PRED
   I=THETA(1)+ETA(1)
   B=THETA(2)+ETA(2)
   A=THETA(3)
   S=THETA(4)
   TNULL=THETA(5)
   Z=A*CAFF*EXP(-(T-1)/TNULL)
   IF (T.EQ.0) THEN
      V=0
      ELSE
      V=T**S
      ENDIF
   MODEL=I+B*V-Z
   Y=MODEL+ERR(1)
$THETA (0.001,5,100) (0.001,10,100) (0.001,4,100) (0.001,1,100)
   (0.001,1,100)
$OMEGA 10 10
$SIGMA 35
$ESTIMATION METH=1
$COVR
```

The "$PROBLEM" statement introduces the nonlinear mixed-model regression analysis to NONMEM, and gives it a name (arbitrarily set to "CAFFEINE"). The "$DATA" statement tells NONMEM where to find the data. In the "$INPUT" statement, the four columns in the data spreadsheet are assigned to the variables ID, T, CAFF, and DV, where the latter stands for "dependent variable" and corresponds to y.

The "$PRED" statement contains the actual regression model. The five fixed effects in the model are automatically handled as a vector THETA with five elements; the two random effects are represented by a vector ETA with two elements (whose parameter estimates ω^2 and λ^2 are in the corresponding vector OMEGA); and the error term ε is a vector ERR with one element (whose parameter estimate σ^2 is in the corresponding vector SIGMA). The model of Eq. (39) is constructed using these building blocks, with I representing I_{ji}, B representing b_{ji}, A corresponding to a, S corresponding to s, and TNULL standing for T_0. The substitute variable V is defined as in the PROC NLMIXED command shown previously. The last line of the "$PRED" statement (which must begin with "Y=") contains the complete model of Eq. (39).

The "$THETA" statement gives the initial values as well as the boundaries for the fixed effects parameters in the THETA vector, in the format "(lower boundary, initial value, upper boundary)"; the "$OMEGA" and "$SIGMA" statements give the initial values for ω^2 and λ^2, and for σ^2, respectively. The "$ESTIMATION" statement specifies details about the numerical procedures to be used, which are beyond the scope of this chapter. The "$COVR" statement, finally, requests computation of the covariance matrix, which is needed for estimation of the standard errors of the parameter estimates.

Table IX shows the parameter estimates resulting from this nonlinear mixed-model regression analysis. The estimate for the time constant T_0 in the model of Eq. (39) is essentially the same as in the model of Eq. (36) (cf. Table VII). However, the estimated standard error of T_0 is increased rather than decreased,[37] suggesting no improvement in this aspect of the model due to the addition of the random effect for b. Indeed, the improvement in the model is nonsignificant overall, as assessed by comparing the model of Eq. (39) (full model) with that of Eq. (36) (reduced model, without parameter λ^2) by means of the likelihood ratio test ($\chi^2[1] = 2.39, p = 0.12$). It follows that among the models we investigated for the data at hand, the preferred model is given by Eq. (36). Further, our best estimate for the half-life of the efficacy of sustained low-dose caffeine in this experiment remains 0.86 days.

Correlation Structures in Mixed-Model Regression Analysis

The statistical power and the efficiency of repeated-measures designs like the present study arise from the correlation of data points within individuals, and the associated distinction of within-subjects variance from between-subjects variance.[2] In most cases, the correlation between the data points within individuals is not a priori known. In practice, therefore, a correlation structure is picked during data analysis, in the hopes that it

---

[37] A threat to the accuracy of standard error estimates is model misspecification (i.e., when the error variance is not distributed normally as specified). In the PROC NLMIXED command in SAS,[24] we have observed cases in which the estimated standard errors were 50% smaller than the true standard errors (as assessed with bootstrap simulations). A correction is available for the covariance matrix from which the standard errors are derived, making them more robust against symmetric nonnormality in the error term. It is commonly referred to as quasi-maximum likelihood (QML) estimation of the covariance matrix. The QML estimate of the covariance matrix is the default output of the "$COVR" statement in NONMEM,[35] but it is not available in PROC NLMIXED in SAS.[24] This may partly explain the increase in the estimated standard error of T_0. See T. Bollerslev and J. M. Wooldridge, *Econometri. Rev.* **11,** 143 (1992).

TABLE IX
PARAMETER ESTIMATES FOR NONLINEAR MIXED-EFFECTS REGRESSION
MODEL WITH TWO RANDOM EFFECTS[a]

Parm	Name	Est	SE
I_0	THETA(1)	4.9	1.1
ω^2	OMEGA(1)	20.3	5.6
b_0	THETA(2)	9.6	3.0
λ^2	OMEGA(2)	6.3	5.0
a	THETA(3)	6.5	3.6
s	THETA(4)	0.50	0.24
T_0	THETA(5)	1.3	2.6
σ^2	SIGMA(1)	26.3	5.6

[a] Results are shown from nonlinear mixed-model regression analysis of psychomotor vigilance performance data over 4 days of sleep deprivation, using the hypothesis-driven model in Eq. (39) with two random effects to assess the duration of the efficacy of sustained low-dose caffeine (relative to placebo). As computed using NONMEM, the table shows the parameters (Parm), their names in the NONMEM macro (Name), their estimates (Est), and their standard errors (SE).

resembles the true correlation structure. There is a rich selection of possible correlation structures (or "covariance structures") that take into account interindividual variability (i.e., random effects) as well as systematic correlations in the residual variance over time.[38] For linear mixed-model regression analysis, methodology for a variety of correlation structures is readily available and implemented in computer software (e.g., in the PROC MIXED command in SAS[25]). For nonlinear mixed-model regression analysis, the implementation of covariance structures is less straightforward. The default situation in nonlinear mixed-model regression is known as the compound symmetry correlation structure,[2] which results from implementing a "variance components" model. A variance components model assumes that the random effects are independent variance components (i.e., having zero covariance).

The variance components model is used for all mixed-model regression analyses in this chapter, including the model of Eq. (36). This model has a random effect on the intercept via the term η of Eq. (25), which has variance ω^2. Although it was not explicitly mentioned in the previous sections, the model actually has a second random effect in the form of the error term ε, which has variance σ^2. These two random effects represent the

[38] R. C. Littell, J. Pendergast, and R. Natarajan, *Stat. Med.* **19**, 1793 (2000).

between-subjects variance and the within-subjects variance, respectively, in this regression model. In a variance components model with a normally distributed random effect on the intercept and normally distributed error variance, the correlation structure is fully determined by the intraclass correlation coefficient (ICC),[39] which is estimated as

$$ICC = \omega^2/(\omega^2 + \sigma^2) \tag{41}$$

In such variance components models the correlation between *each pair* of data points of a given subject is assumed to be equal to the ICC. For the nonlinear mixed-effects regression model of Eq. (36), this correlation can be estimated using the results in Table VII:

$$ICC = 28.6/(28.6 + 29.2) = 0.49 \tag{42}$$

which means that the correlation between each pair of data points within each subject is (implicitly) assumed to be 49%.[40]

It is noteworthy that the value of the ICC has an additional, complementary interpretation.[41] By definition of Eq. (41), the ICC expresses the between-subjects variance as a fraction of the total variance not explained by the fixed effects in the model.[42] In Eq. (36), therefore, the ICC quantifies the importance of systematic interindividual differences in the intercept with respect to overall variability in the data around the regression model. Using this interpretation of the ICC, studies repeating sleep deprivation in the same individuals have revealed systematic interindividual differences in performance deficits resulting from sleep deprivation, with ICC values greater than 0.5.[41] This underlines the importance of taking

[39] If the random effect is not on the intercept but on another component of the model, computation of the ICC is ambiguous. For instance, suppose that the model in Eq. (39) would not have a random effect on the intercept, leaving only the random effects represented by β and the error term ε. The value and the unit of the variance λ^2 for β would depend on the magnitude (and unit) of the factor t^s in that model (e.g., whether t is expressed in days or in hours). This would cause obvious problems for the computation of the ICC, which do not arise if the random effects are on the intercept and the (additive) error term only (and the distribution of both random effects is assumed to be normal).

[40] Investigating whether or not this correlation structure is realistic given the data at hand is beyond the scope of this chapter. The data can be analyzed repeatedly with different correlation structures and the results compared by means of a statistical information criterion to select a correlation structure that best fits the data. This is often done, for instance, when an autoregressive structure is suspected but the order of autoregression is yet to be determined.

[41] H. P. A. Van Dongen, G. Maislin, and D. F. Dinges, *Aviat. Space Environ. Med.* (Suppl.), in press.

[42] For more about explained variance in mixed-model regression analysis, see T. A. B. Snijders and R. J. Bosker, *Sociol. Methods Res.* **22,** 342 (1994).

such interindividual differences into account (e.g., with random effects) when modeling data from sleep deprivation experiments.

Conclusion

In this chapter, we considered the analysis of longitudinal data in the presence of interindividual differences. We first described repeated-measures analysis of variance (ANOVA), a traditional technique for the analysis of longitudinal data tailored to the comparison of the means of subsets of the data. We showed that this technique is not robust to systematic interindividual differences. We then discussed linear mixed-model regression analysis. We employed this technique to mimic repeated-measures ANOVA while adding robustness against systematic interindividual variability (i.e., mixed-model ANOVA). This application of mixed-effects modeling is especially useful if no a priori expectations exist about the shape of the data's temporal profile. For hypothesis-driven analysis of time series data, however, mixed-effects models frequently involve nonlinearity in the parameters. Therefore, we also considered nonlinear mixed-model regression analysis. Our aim was to convey a basic understanding of the mathematical and statistical issues involved in mixed-model regression analysis. For this purpose, we included specific examples for how to implement mixed-effects regression models in computer software (i.e., SAS[24] and NONMEM[35]). In the process, we assessed the duration of the efficacy of sustained low-dose caffeine during an experiment involving 88 h of continuous wakefulness. The data from this study are publicly available (see Table I); thus, the findings reported here can be replicated as an exercise for familiarizing oneself with mixed-effects modeling.

As the generic mechanisms of many physiological and pharmacological phenomena are better understood, there will be a growing—and much needed—interest in interindividual differences to explain the diversity in the parameters of these mechanisms within populations. As a consequence, there will be an increasing demand for data analysis techniques capable of dealing with interindividual differences. With the major enhancements of computer power seen in recent years, mixed-model regression analysis has become feasible on common personal computer platforms. We expect, therefore, that mixed-model regression will become a standard by which longitudinal data are analyzed in the twenty-first century. We hope that the present chapter will facilitate this trend.

[11] Sample Entropy

By Joshua S. Richman, Douglas E. Lake, and
J. Randall Moorman

Introduction

The findings of deterministic dynamics in seemingly random physical process have excited biological researchers who collect time series data. The best tools for this kind of analysis require extremely long and noise-free data sets that are not available from biological experiments. Nonetheless, one such tool has found widespread use. In 1991, Pincus adapted the notion of "entropy" for real-world use.[1] In this context, entropy means order or regularity or complexity, and has roots in the works of Shannon, Kolmogorov, Sinai, Eckmann and Ruelle, and Grassberger and co-workers. The idea is that time series with repeating elements arise from more ordered systems, and would be reasonably characterized by a low value of entropy. Were the data sets infinite and perfect, it would be possible to determine a precise value of entropy. Biological data sets are neither, but Pincus had the important insight that even an imperfect estimate of entropy could be used to rank sets of time series in their hierarchy of order. He introduced approximate entropy (ApEn), and many papers have appeared drawing conclusions about the relative order of physiological processes.

In principle, the calculation is simple enough, and is shown schematically in Fig. 1. ApEn quantifies the negative natural logarithm of the conditional probability (CP) that a short epoch of data, or template, is repeated during the time series. Having selected a template of length m points, one identifies other templates that are arbitrarily similar and determines which of these remain arbitrarily similar for the next, or $m + 1$st point. "Arbitrarily similar" means that points are within a tolerance r of each other, where r is usually selected as a factor of the standard deviation (SD). The negative logarithm of the conditional probability is calculated for each possible template and the results averaged. If the data are ordered, then templates that are similar for m points are often similar for $m + 1$ points, CP approaches 1, and the negative logarithm and entropy approach 0.

The concepts are solid and the potential utility is great. We found, however, that there are practical issues of great importance in implementing the algorithm. These findings motivated us to develop sample entropy

[1] S. M. Pincus, *Proc. Natl. Acad. Sci. USA* **88**, 2297 (1991).

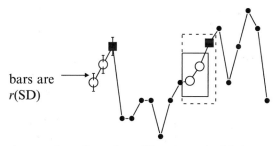

bars are ⟶
r(SD)

■ A_i = number of matches of length $m+1$ with ith template

○ B_i = number of matches of length m with ith template

$$ApEn \approx \Sigma - \log\ (1 + A_i)/(1 + B_i)$$
$$SampEn = -\log\ ((\Sigma A_i)/(\Sigma B_i)) = -\log\ A/B$$

For regular, repeating data, A/B nears 1 and entropy nears 0.

FIG. 1. Schematic demonstration of entropy estimation using approximate entropy (ApEn) and sample entropy (SampEn). The time series begins with the ith template. In this example, m is 2. The tolerance for accepting matches is r times the standard deviation, and is shown by the error bars. Here, the template is matched by the 11 and 12th points (solid box), and the $m + 1$st points also match (dashed box). Thus quantities A and B both increment by 1.

(SampEn) as an alternative method for entropy estimation in real world data. In this chapter, we first overview the problems of ApEn and how SampEn addresses them. We next present a formal implementation of SampEn and discuss the practical issues of optimization of parameters and data filtering. This is followed by a discussion of the difficulties with short data sets and nonstationary data. We end with comments on interpretation of entropy estimates and a direct comparison of ApEn and SampEn. The algorithms discussed are available at www.Physionet.org. For full details, we refer the reader to our original papers.[2,3]

Motivation for SampEn Analysis

In our initial implementation of ApEn analysis of heart rate dynamics, we encountered practical questions.

1. What if some templates have no matches, and the CPs are not defined? Pincus follows the teaching of Eckmann and Ruelle and allows

[2] D. E. Lake, J. S. Richman, M. P. Griffin, and J. R. Moorman, *Am. J. Physiol.* **283,** R789 (2002).
[3] J. S. Richman and J. R. Moorman, *Am. J. Physiol.* **278,** H2039 (2000).

templates to match themselves. Thus if there are no other matches, the CP is 1 and ApEn is 0, a report of perfect order. If there are only a few template matches, then the result is biased toward 0, and the bias resolves with lengthening data sets and more template matches. Pincus and co-workers have explicitly described the bias of ApEn and have contended that the important goal of reporting a correct hierarchy of order is preserved, a feature named "relative consistency." Thus if time series A arises from a more ordered system than time series B, then ApEn of A will always be less than ApEn of B regardless of the number of template matches and the extent of the bias.

We have developed SampEn statistics to reduce the bias of ApEn statistics.[2,3] We have found that SampEn preserves relative consistency more often than ApEn.

2. How long should a template be, and how similar should "arbitrarily similar" be? That is, how does one pick m and r? The usual suggestion is that m should be 1 or 2, noting that there are more template matches and thus less bias for $m = 1$, but that $m = 2$ (or greater) reveals more of the dynamics of the data. The usual suggestion is that r should be 0.2 times the SD of the data set on empirical grounds.

We have proposed a systematic approach to selecting m and r based on evaluation of novel error metrics for SampEn that are discussed later.

3. Does a low value of entropy always mean increased order?

No differential diagnosis has been suggested for ApEn. We have found that time series with spikes have low values of ApEn and SampEn, a direct consequence of the practice of basing the tolerance r on the SD.[2] Spikes inflate the SD and allow many template matches of lengths m and $m + 1$ in the baseline. The high CP leads inevitably to a low value for the entropy, but it is not intuitively correct that a large number of matching templates in the baseline necessarily reflects order.

Sample Entropy Calculation

As a statistic, $SampEn(m,r,N)$ depends on three parameters. The first, m, determines the length of vectors to be considered in the analysis. That is, given N data points $\{u(j): 1 \leq j \leq N\}$, form the $N-m+1$ vectors $x_m(i)$ for $\{i \mid 1 \leq i \leq N - m + 1\}$ where $x_m(i) = \{u(i + k): 0 \leq k \leq m - 1\}$ is the vector of m data points from $u(i)$ to $u(i + m - 1)$. The distance between two vectors, denoted $d[x_m(i), x_m(k)]$, is defined to be max $\{|u(i + j) - u(k + j)|: 0 \leq j \leq m - 1\}$, the maximum difference between their corresponding scalar components.

The original formulation of SampEn closely followed the Grassberger–Procaccia correlation integral. However, a more intuitive and less notationally intensive approach simply considers $SampEn(m,r,N)$ to be the negative natural logarithm of the empirical probability that $d[x_{m+1}(i), x_{m+1}(k)] \leq r$ given that $d[x_m(i), x_m(k)] \leq r$. Where the values of the parameters are specified let B denote the number of pairs $x_m(i)$, $x_m(k)$ such that $d[x_m(i), x_m(k)] \leq r$, and let A then be the number of pairs of vectors $x_{m+1}(i)$, $x_{m+1}(k)$ such that $d[x_{m+1}(i), x_{m+1}(k)] \leq r$. Then $SampEn(m,r,N) = -\ln(A/B)$. For simplicity, we refer to a match of two vectors of length m as a *template match,* and the matching of two vectors of length $m + 1$ as a *forward match.*

The computationally intensive aspect of the algorithm is simply counting the numbers A and B, which at least requires checking the distance between the $[N(N - 1)]/2$ pairs of points, then counting the number of vectors that match for m and $m + 1$ points. There are two additional computational considerations, both of which are fully discussed in the initial paper by Richman and Moorman.[3] First, we do not compare any vector with itself since this provides no new information. Second, although the vector $x_m(N - m + 1)$ exists, we do not use it for comparisons, since the vector $x_{m+1}(N - m + 1)$ is not defined. That is, if some $x_m(i)$ were to match $x_m(N - m + 1)$, they could not both be incremented, and thus could contribute only to B but not A. In practice, for a data set of reasonable size this will have little effect, but scenarios can be constructed where it could significantly alter results.

While SampEn is often used with just one fixed value of m, there is an implementation of the algorithm that efficiently calculates $SampEn(k,r,N)$ for all k from 1 up to m. The basic idea is to build up runs of points matching within the tolerance r until there is not a match and keep track of template matches $A(k)$ and $B(k)$ of all lengths k up to some specified parameter m. If a particular run ends up being of length 4, for example, then that means that 1 is added to the count for template matches of length 4. In addition, there are 2 template matches of length 3, 3 of length 2, and 4 of length 1 that need to be added to the corresponding counts. A special distinction is needed when a run ends at the last point in the data where the $A(k)$ counters are incremented but the $B(k)$ counters are not.

Practical Issues with SampEn Calculation

Optimizing Parameters

Having decided how to manage the data the next task must be to optimize the parameters for $SampEn(m,r,N)$ by some rational strategy. For most current applications the parameters must be fixed to allow valid comparisons. Circumstances that may indicate varying parameters will be discussed separately. The parameter N is usually taken as the size of the data set. Since N determines the range of $SampEn(m,r,N)$ by providing an upper limit for B, care must be used when comparing epochs of differing lengths, especially if $SampEn(m,r,N)$ approaches its theoretical maximum for any epoch. Selecting m and r is more involved. Assuming that there is some underlying deterministic structure, as m grows larger, the conditional probability of a forward match should increase, since conditioning on a longer vector should increase predictive accuracy. On the other hand, due to the noise from measurement errors or system interaction the number of template matches will decrease as m increases simply because it is less likely that the noise level will permit long templates to match at all. In fact, beyond a certain length the templates will largely be due to chance rather than dynamic similarity. In addition, and likely before this point is reached, the matches will be so few that the statistics become unreliable. Conversely, when m is too small, template matches are plentiful (B is large), but not enough predictive information is contained in the short template leading to an underestimation of the probability of a forward match. The compromises involved in choosing r are similar. An r that is too small, smaller than the typical noise amplitude, will result in many vectors that are actually similar failing to match. On the other hand, when r is too large, $SampEn(m,r,n)$ loses its discriminating power entirely since most templates will look similar to one another given sufficiently lax matching conditions.

The ideal choice then would be to make m as large and r as small as possible, while ensuring that B remains large enough to ensure precise statistics. Ideally, one could choose m based upon knowledge of the time scale of the underlying process, and r based upon knowledge of the scale of signal noise. In practice, we have chosen m by fitting an autoregressive model to the data and setting m to be the optimal order of the model. This method has the advantage of drawing upon a vast literature and established methods for optimizing such models. We then choose r to minimize the relative error of A/B and $SampEn(m,r,N)$.

Given the B template matches calculating the standard error of the conditional probability estimate $CP = A/B$ is nontrivial. This is largely due to the fact that vector comparisons are not independent. Not only should there be some dependence due to the underlying process but some comparisons are formally dependent on others because they overlap and thus share data points. Nevertheless an approximation of this standard error is given by

$$\sigma_{CP}^2 = \frac{CP(1 - CP)}{B} + \frac{1}{B^2}\left[K_A - K_B(CP)^2\right]$$

where K_A denotes the number of overlapping pairs or $m + 1$ point vectors, and K_B denotes the number of overlapping pairs of m point templates. That is, K_A is the number of pairs $[x_{m+1}(i), x_{m+1}(j)]$, $[x_{m+1}(k), x_{m+1}(l)]$ in which a vector from one pair shares at least one data point with a vector from the second pair and both pairs match within r. This would be the case if, for example, $x_{m+1}(i)$ and $x_{m+1}(k)$ had a point in common. K_B is defined analogously for vectors of length m. Counting K_A and K_B requires both care and a considerable amount of computer time. From this we can approximate the standard error of $SampEn(m,r,n)$ by σ_{CP}/CP. Having chosen m we then advise choosing r to minimize the larger of σ_{CP}/CP and $\sigma_{CP}/{-}\log(CP)CP$, the relative errors of the empirical CP and $SampEn(m,r,n)$, respectively. The estimate of σ_{CP} is also useful in that it allows for the calculation of approximately 95% confidence intervals for $SampEn(m,r,n)$ by $1.96(\sigma_{CP}/CP)$.

At this point another consideration must be addressed regarding r. It is standard practice to set r to be some multiple of the standard deviation of the data. This effectively normalizes the data, adjusting for differences of scale. This is appropriate if the analysis is driven by the search for order in the dynamics. If, however, the goal is to efficiently distinguish various data sets the rescaling can make two data sets appear identical when they clearly are not. There can also be problems with interpretation since broad trends or transient phenomena can inflate the variance to the point where all local dynamics appear similar only because they have made the tolerance r too coarse to discern genuine similarity.

Having chosen a method for selecting parameters it is then necessary to choose either some or all of the data and calculate $SampEn(m,r,N)$ for a range of the parameters. Since these calculations can be time consuming it is often advisable to select either a random sample from all the data or some characteristic data sets to find parameters that are broadly appropriate for all the data. This is not, of course, a definitive method for choosing parameters, but one that balances the requirements of discerning distinctive features of the data (sufficiently large m and small r) and allowing for accurate and precise estimates (adequate B).

Other suggested approaches involve choosing the parameters that maximize the deviation of $SampEn(m,r,N)$ from $SampEn(0,r,N)$, the negative logarithm of the probability that a pair of points will match within r given no other information, i.e., the value of $SampEn(m,r,N)$ if the data were in random order. This method focuses on locating the parameters that detect the most "order" in the data. Above all, the most important factor is choosing parameters appropriate for the data at hand. It cannot be overemphasized that prior to any meaningful analysis the parameters must be chosen to be suitable for the data. We have no reason to suspect that there are universally optimal parameters suitable for any and all data sets. We therefore advocate that they be chosen either by optimizing their empirical properties on a subset of the data, or by choosing reasonable parameters based on prior knowledge and checking their performance by, for example, making sure that the resulting confidence intervals are not too wide.

Another strategy for distinguishing data sets is to view the identification of optimal parameters from another angle. We have so far considered identifying the finding parameters that maximize the usefulness of SampEn statistics as the characteristic measure of the time series. We could choose instead to optimize properties of $SampEn(m,r,N)$ and use the resulting parameters to characterize the time series. This approach has the potential advantage of providing a categorical classification by m of predictive scale (or by a time delay τ, if using the time delay techniques discussed later) or a continuous measure of structural detail by r. Again, a reasonable approach is to maximize the difference between $SampEn(m,r,N)$ and $SampEn(0,r,N)$. Another would be to maximize the difference between $SampEn(m,r,N)$ for the time series and for a surrogate time series incorporating the linear correlation structure of the time series.[4]

Once the parameters have been chosen all the data may be analyzed and the values of $SampEn(m,r,N)$ incorporated into regression models or tested via their confidence intervals as to whether they differ significantly from one another. Again, under these conditions SampEn statistics appear to provide a robust way to discriminate between time series. More care is needed to say what proportion of the measure is due to order and what is due to nonstationarity.

Filtering

Many time series techniques assume at least weak stationarity, so data are often subjected to filtering to render them approximately stationary prior to analysis. This alters the correlation structure of the data in a way that can change the SampEn analysis, and should be done judiciously. The accuracy and precision of $SampEn(m,r,N) = -\ln(A/B)$ is limited by the magnitude of B. Trends in the data can separate points whose local dynamics are similar but whose locations are far removed. When trends are understood, they may be removed, laying bare the unexplained dynamics. Nevertheless, there is always the risk that the most interesting features could be blunted or obliterated by filtering and smoothing. Leaving the data unfiltered removes some danger of manufacturing spurious matches at the expense of lowering B. Fortunately, only large data sets really have

[4] T. Schreiber and A. Schmitz, *Physica. D* **142,** 346 (2000).

enough points to display complicated trends, and these are the least likely, by virtue only of their size, to be hampered by a low value of B.

This discussion has so far assumed that the time scale of the studied dynamics are near to the scale of sampling. If, however, one is more interested in longer time scales, that is, when the data are effectively oversampled, the usual approach can be modified in several ways. These generally involve changing or altering the reconstruction vectors $x_m(i)$ to include time delays. This is accomplished by using vectors of the form $x_m^\tau(i) = \{u(i + \tau k): 0 \leq k \leq m - 1\}$ for a time delay τ. Techniques for choosing τ are discussed extensively in the literature.[5] In this case there is little reason to use every possible vector since for an oversampled process $x_m^\tau(i)$ will match $x_m^\tau(i + 1)$ as often and uninterestingly as $x_m(i)$ will match $x_m(i + 1)$. Thus some consideration needs to be given as to which subsample of vectors should be used for analysis. Recently Costa et al.[6] developed multiscale entropy analysis (MSE), which compresses the data by constructing coarse time series of the form $y^\tau(j) = 1/\tau \sum_{i=(j-1)\tau+1}^{j\tau} u(i)$ for $1 \leq j \leq N/\tau$, that is, by dividing the series into a sequence of non-overlapping windows of τ points and representing the window by its sample mean. The new point $y^\tau(j)$ corresponds to the mean of the jth window of τ points.

SampEn analysis was then carried out on these new series to give information regarding the dynamics on larger time scales. Clearly this is easily generalized to a family of such techniques. A more general version less sensitive to the window locations would be to replace the series $\{u(j): 1 \leq j \leq N\}$ by a moving averaged series $\{u^\tau(j): \tau \leq j \leq N\}$ where each point is the mean of the previous τ points of the original time series. SampEn analysis could then be carried out on the delayed vectors $\tilde{x}_m^\tau(i) = u(i + k\tau): 0 \leq k \leq m - 1\}$. Doing this for various values of τ would broaden the MSE approach. It may also be desirable to utilize weighted averaging rather than the simple window average depending on the desired smoothing properties.

If the data are fairly precise and merely oversampled, it may be worthwhile to use usual m-vectors and $m + 1$ vectors of the form

$$x_{m+1,\tau}(i) = \{u(i + k\tau), \quad u[i + (m - 2) + \tau]: 0 \leq k \leq m - 2\}$$

Thus the first m points of the vector are the usual m-dimensional reconstruction vector while the last point has been sampled at the delay τ. This allows the oversampled (and dynamically uninteresting) portion of the data to fix the system's location in phase space more precisely, while the

[5] H. D. I. Abarbanel, "Analysis of Observed Chaotic Data." Springer-Verlag, New York, 1996.
[6] M. Costa, A. L. Goldberger, and C. K. Peng, Phys. Rev. Lett. **89**, 068102 (2002).

comparisons at the next τth point allows the system enough time to evolve so that matches will be meaningful. Experimentation will be required to select the optimum choice of τ.

SampEn analysis does not require that any assumptions be made regarding the stationarity of the data. However, in the absence of approximate stationarity care must be taken with the interpretation of SampEn statistics. Indeed, as we shall see throughout this discussion, care must be taken in all circumstances when interpreting SampEn statistics as indicative of order in the data.

Difficult Data

Short Data Sets

The analysis of very short data sets may call for extra care. $SampEn(m,r,N)$ is bounded by 0 and $\ln [(N - m)(N - m - 1)]/2$. Thus short data sets will have a decreased range. They will also have smaller values of B and thus less precise statistics. This will be exacerbated by the fact that in a small data set a higher proportion of comparisons involves overlapping templates, a factor that will tend to inflate the variance of $SampEn(m,r,N)$. We have also shown previously that for short sets of i.i.d. random numbers, overlapping templates lead to an average underestimation of A/B. Scenarios can be imagined wherein for oversampled data, the underlying process does not have much time to evolve along any particular template. In these cases overlapping templates will match each other with high probability. Unfortunately, the high estimated CP will be due to the local similarity due to slow evolution of the system rather than self-similarity in phase space.

One obvious remedy would be to disregard comparisons involving overlapping $m + 1$ vectors. There are generally $[(N - m)(N - m - 1)]/2$ distinct pairs of templates. Discarding those pairs of vectors that would overlap when incremented to length $m + 1$, there are $[(N - 2m)(N - 2m - 1)]/2$ distinct pairs. The fraction of matches discarded is then

$$\frac{mN\left(2 - 3\frac{m}{N} - \frac{1}{N}\right)}{N^2\left(1 - \frac{m}{N}\right)\left(1 - \frac{(m+1)}{N}\right)} \cong \frac{2m}{N}$$

for $m << N$. Thus for large N in exchange for a modest reduction in the number of comparisons, we can ensure that the vectors are at least formally, if not dynamically, independent, which should lead to a reduction in SampEn's variance as well as increased confidence that the CP is correctly estimated.

Nonstationary Data

We noted previously that there is no requirement that the data be stationary. The main concern is that the nonstationarity of the data could inflate the tolerance parameter *r*, and thereby induce spurious matches. Given large amounts of data, choosing *r* to be as small as possible should provide a reasonable safeguard. Nevertheless, particularly when confronted with nonstationary data one may decide to average the values of *SampEn(m,r,n)* for windows of length *n < N*, or adopt some other resampling approach. *SampEn(m,r,n)* provides a global measure of the probabilistic self-similarity of the data. Since the hallmark of nonstationarity is a varying probabilistic structure, there may be cases in which it is advisable to perform a piecewise analysis, dividing the time series into sections with homogeneous structure and calculating SampEn for each segment. One could then report the average of the individual SampEn statistics, where each is weighted according to their proportion of the entire series.

Interpretation of SampEn

SampEn was originally intended as a measure of the order in a time series. We have noted, however, that low SampEn statistics, indicative of high CP estimates, cannot be assumed to imply a high degree of order. There are in general two distinct mechanisms for generating high CP estimates. The first is that genuine order has been detected. The second derives from the fact that *r* is usually taken as a proportion of the standard deviation of the series, thus rendering the analysis scale free. When nonstationary features, especially transient "spikes," inflate the variance and thus coarsen the criterion for matching it can happen that virtually all recorded matches are similar only because their dynamic scale is dwarfed by the spikes. This has two ramifications. First, when the aim is to quantify order with SampEn statistics, additional scrutiny is required to ascertain how much of the statistic's value is, in fact, due to order. Second, when the goal is simply to numerically distinguish between data sets, SampEn is very adept at detecting such spikes and generally discerning which epochs are atypical.

When order detection is desired we suggest comparing *SampEn(m,r,N)* to *SampEn(0,r,N)* to see whether conditioning on the templates significantly increased the probability of a forward match. If *SampEn(0,r,N)* lies outside the confidence interval for *SampEn(m,r,N)*, then we are reasonably confident that a significant amount of order was detected. If the goal remains chiefly to discriminate among series, but it is desirable to lean toward detecting order, *r* can be taken to be a proportion of the mean (or variance) of the series of first differences, that is, of the average change from one point to the next. This effectively rescales the series to have similar dynamic scales, and will generally give less weight to spikes. One could of course simply perform SampEn analysis on the series of first differences as well, but this acts as a high-pass filter and will obscure long-range dynamics. Another approach would be to use r as an absolute number, thereby allowing SampEn to distinguish between series that differ only in their scaling. Unfortunately when the scales of the time series vary widely, no r may be suitable to provide comparisons across all sets with a meaningful interpretation for the statistic per se. However, the SampEn statistics will likely still efficiently detect which data sets are atypical.

ApEn and SampEn

We now summarize the differences between sample entropy and approximate entropy and discuss possible bridges between the two approaches. Let B_i denote the number of template matches with $x_m(i)$ and A_i denote the number of template matches with $x_{m+1}(i)$. The number $p_i = A_i/B_i$ is an estimate of the conditional probability that the point x_{j+m} is within r of x_{i+m-1} given that $x_m(j)$ matches $x_m(i)$. ApEn is calculated by

$$ApEn(m,r,N) = -\frac{1}{N-m} \sum_{i=1}^{N-m} \log\left(\frac{A_i}{B_i}\right)$$

and is the negative average natural logarithm of this conditional probability. Self-matches are included in the ApEn algorithm to avoid the $p_i = 0/0$ indeterminate form, but this convention leads to a conditional probability estimate of 1. This necessarily overestimates the true value and leads to a noticeable bias especially for smaller N and larger m. In contrast to the above, SampEn is calculated by

$$SampEn(m,r,N) = -\log\left(\sum_{i=1}^{N-m} A_i \bigg/ \sum_{i=1}^{N-m} B_i\right) = -\log(A/B)$$

which is just negative the logarithm of an estimate of the conditional probability CP of a match of length $m + 1$ given a match of length m.

There are advantages in using SampEn over ApEn. ApEn is more sensitive to bias from short time series and to problems arising from outliers. We discovered that ApEn could give misleading or contradictory results, particularly for very short or noisy time series. Sample entropy was developed to improve some of these properties while maintaining the spirit of measuring the probability that two vectors that are close for m points remain close. In all cases studied so far SampEn appears to be a more robust and less biased statistic than ApEn. SampEn's formulation is more amenable to the construction of confidence intervals that give approximate guidelines for significance tests.

Despite the advantages of using SampEn, there may be theoretical and personal preference issues that necessitate using ApEn. ApEn statistics are much more closely related to Shannon's definition of information-theoretic entropy that spurred the field of information theory more than 50 years ago and they possess an additive property that is not rigorously shared by sample entropy. Rukhin[7] gives a clear statement of the relationship between Shannon's entropy and ApEn. In addition, it is formally very closely related to the entropy of statistical mechanics. There exists a vast literature exploring the properties of such functions from both perspectives. In particular, Rukhin has recently proven that for discrete series, ApEn statistics are asymptotically distributed as a χ^2 random variable. This suggests a future practice of applying ApEn to discretized signals. The trade-off here is that while information is lost to discretization, the parameter r can be discarded since sequences of discrete values can be considered to match either exactly or not at all. We note that while there are, as yet, no theoretical proofs of SampEn's asymptotic properties, Monte Carlo studies have given no reason to suppose that SampEn's asymptotic properties are not well behaved.

Approximate entropy also is optimally suited to measure the Gaussianity of a distribution and a process. Berg's theorem established that the maximum entropy for a random process with finite variance is attained by a Gaussian process. Thus ApEn values departing from the theoretical maximum indicate a lack of Gaussianity. SampEn can also be effectively used as a measure of Gaussianity though its maximum occurs for non-Gaussian random processes. When ApEn is used, modifications of how ApEn handles zero and small number of matches can help minimize its bias and begin to approach the statistical stability of sample entropy. This is an important open area for research. One of the major sources of ApEn bias results when a template does not match any others, whereupon ApEn estimates a conditional probability of 1. Pincus has suggested correcting this by incorporating a factor ε into calculation, thus setting the CP estimate in the absence of matches.

In any implementation of ApEn, we advocate not counting self-matches. When $A_i = B_i = 0$ (or near 0), the estimate A_i/B_i can be replaced with some value ε that is sufficiently smaller than 1 to indicate the likely

[7] A. L. Rukhin, *J. Appl. Prob.* **37,** 88 (2000).

uncertainty of matching if no matching templates exist.[6,8,9] The value cannot be too small or $-\log(\varepsilon)$ is prohibitively large and can lead to an upward bias. Clearly, selecting the right value of ε is dependent on the process being studied. One approach is to use estimates $ApEn(k,r,N)$ with $k <$ m based on shorter templates (and thus less biased) to approximate $-\log(A_i/B_i)$ when matches are lacking. This is a generalization of the method used by Porta and co-workers[6,9] in the calculation of *corrected conditional entropy* (CCE) where the Shannon entropy of the process is used to approximate $-\log(A_i/B_i)$ when $B_i = 0$. We suggest that when this strategy is employed that the ε factor should correct toward the SampEn CP estimate. This novel and promising approach would take advantage of the statistical value of sample entropy by setting $\varepsilon = CP = A/B$ from the SampEn calculation and lead to a hybrid of the two algorithms.

There remain several unresolved issues in the use of SampEn. Much work needs to be done to evaluate and rigorously establish the statistical properties of SampEn statistics. Most desirable would be proof of the statistics' asymptotic distributions. It will also be important to develop tests for the proportional contributions of order and nonstationarity. Studies also need to be undertaken to ascertain how the various optimization methods lead to the selection of parameters, and how sensitive those selections are to outliers in the data. Given the parameter dependence of the statistics and its relation to comparing datasets, we are also working to develop related parameter-free statistics.

Thus we conclude the following:

1. A low value of ApEn is due to bias, order, or spikes. The relative contributions cannot be quantified.
2. A low value of SampEn is due to order or spikes.

Our practice is to use SampEn with optimized choices of m and r. Work remains on confident parsing of the results to order and spikes.

[8] S. M. Pincus and A. L. Goldberger, *Am. J. Physiol.* **266,** H1643 (1994).
[9] A. Porta, G. Baselli, F. Lombardi, N. Montano, A. Malliani, and S. Cerutti, *Biol. Cybern.* **81,** 119 (1999).

[12] Calculating Sedimentation Coefficient Distributions by Direct Modeling of Sedimentation Velocity Concentration Profiles

By JULIE DAM and PETER SCHUCK

Introduction

Sedimentation velocity analytical ultracentrifugation can give rich information about the purity, molar mass, state of association, protein interactions, hydrodynamic shapes, conformational changes, and size distributions, among other properties of proteins.[1–3] It is based on a conceptually very simple principle of applying a gravitational force to the protein solution and observing the resulting changes in the concentration distribution.[4,5] Sedimentation velocity has been continuously refined during the past 80 years, and historically has provided many important contributions to the development of both polymer chemistry and biochemistry, starting with the discovery of proteins being macromolecules and of well-defined sizes.[6] Most of the theoretical foundations have long been laid, such as in 1929 by Lamm[7] in Svedberg's laboratory, the partial differential equation for macromolecular sedimentation and diffusion in the centrifugal field. However, the past several years have witnessed important developments, as it has become possible to solve the Lamm equation on laboratory computers and to use it now routinely to model experimental data from increasingly complex systems.[8–16] This is in confluence with more precise and substantially larger

[1] J. Lebowitz, M. S. Lewis, and P. Schuck, *Protein Sci.* **11,** 2067 (2002).
[2] G. Rivas, W. Stafford, and A. P. Minton, *Methods Companion Methods Enzymol.* **19,** 194 (1999).
[3] T. M. Laue and W. F. I. Stafford, *Annu. Rev. Biophys. Biomol. Struct.* **28,** 75 (1999).
[4] T. Svedberg and K. O. Pedersen, "The Ultracentrifuge." Oxford University Press, London, 1940.
[5] H. K. Schachman, "Ultracentrifugation in Biochemistry." Academic Press, New York, 1959.
[6] B. Elzen, "Scientists and Rotors: The Development of Biochemical Ultracentrifuges." Dissertation, University Twente, Enschede, Netherlands, 1988.
[7] O. Lamm, *Ark. Mat. Astr. Fys.* **21B**(2), 1 (1929).
[8] J. S. Philo, *Biophys. J.* **72,** 435 (1997).
[9] P. Schuck, C. E. MacPhee, and G. J. Howlett, *Biophys. J.* **74,** 466 (1998).
[10] P. Schuck and D. B. Millar, *Anal. Biochem.* **259,** 48 (1998).
[11] P. Schuck, *Biophys. J.* **75,** 1503 (1998).
[12] W. F. Stafford, *Biophys. J.* **74**(2), A301 (1998).
[13] B. Demeler, J. Behlke, and O. Ristau, *Methods Enzymol.* **321,** 36 (2000).

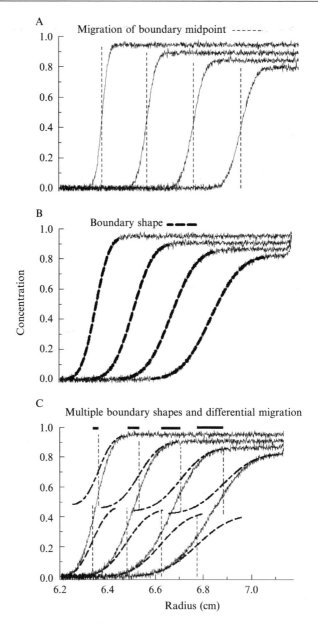

FIG. 1. Schematic principle of different strategies for sedimentation velocity data analysis. The concentration profiles at different times are the measured data (solid lines). (A) The determination of the boundary midpoints (dashed vertical lines) and the migration of the midpoint with time. If the midpoint is determined by the second moment methods,[5,17] the result

experimental data basis, provided, for example, by the laser interferometry detection system of the commercial analytical ultracentrifuges.[18]

One of the key problems in the interpretation of the macromolecular concentration distributions and their evolution is the separation of the effects of diffusion and sedimentation. This is particularly important in the study of proteins, where frequently the diffusional transport equals or exceeds the migration caused by the centrifugal force. This problem can be illustrated by considering the sedimentation profiles of a sample of a non-interacting protein species (Fig. 1). Diffusion effects can be experimentally minimized by applying a high centrifugal field to increase the sedimentation rate so as to form a sharp sedimentation boundary, and an average sedimentation rate of the sample can be determined by simply measuring the velocity of this boundary (Fig. 1A). However, it is clear that this neglects an enormous amount of information contained in the shape of the evolving concentration distributions. It also does not permit the analysis of molar mass, and the characterization of different protein subpopulations. The simplest approach for taking diffusion into account is to interpret the spread of the boundary shape as if resulting from the diffusion of a single species (Fig. 1B). In principle this permits the determination of the diffusion coefficient and the molar mass of the sedimenting species. However, in practice this approach frequently fails, because it requires the protein under study to be highly homogeneous. For heterogeneous mixtures where the root-mean-square displacement from diffusion is in the same

---

is the weight-average sedimentation coefficient. (B) In addition to the displacement of the boundary midpoint, the spread of the boundary can be interpreted to result from macromolecular diffusion, and modeled with a Lamm equation solution (or an approximation thereof) (dashed line). If the sample under study consists of a single species, the resulting diffusion coefficient allows determination of the molar mass of the macromolecule. (C) For a two-component system, however, an additional boundary spreading can be caused by the differential migration of the sedimenting species (indicated by horizontal bars). For proteins, frequently, the separation of the species does not exceed their diffusional spread, causing a complex boundary shape. An "apparent diffusion," if evaluated from the overall boundary shape, is not a meaningful parameter, and the existence of multiple species may not be obvious. The quantitative analysis of the data in (C) requires distributions of Lamm equation solutions.

---

[14] J. Behlke and O. Ristau, *Biophys. Chem.* **95,** 59 (2002).

[15] P. Schuck, *Biophys. Chem.* in press (2004).

[16] P. Schuck, *Biophys. Chem.* in press (2004).

[17] P. Schuck, *Anal. Biochem.* **320,** 104 (2003).

[18] D. A. Yphantis, J. W. Lary, W. F. Stafford, S. Liu, P. H. Olsen, D. B. Hayes, T. P. Moody, T. M. Ridgeway, D. A. Lyons, and T. M. Laue, *in* "Modern Analytical Ultracentrifugation" (T. M. Schuster and T. M. Laue, eds.), p. 209. Birkhäuser, Boston, 1994.

order as the separation of species due to their different sedimentation rate (Fig. 1C), such as mixtures of small or similar-sized proteins, a complex evolution of boundary shape and boundary location is observed in which the boundary spreading is due to the combined effects of diffusion and differential migration. In this situation, all known data transformations and extrapolation techniques usually fail to provide either a good description of diffusion properties of the mixture or a resolution of the sedimentation coefficients of the different species.[19]

Modern sedimentation velocity methods using solutions of the Lamm equation assume a specific model for macromolecules under study, calculate their sedimentation behavior, and fit the model to the experimental data by least-squares techniques. This chapter focuses on the currently most general direct boundary model, which is that of a continuous distribution of noninteracting proteins. In the form of a sedimentation coefficient distribution $c(s)$,[20] it addresses the problem outlined previously and provides a robust estimation of the populations of species with different sedimentation rates, at a high resolution, and along with a description of the diffusion of the ensemble via a weight-average frictional ratio.[19] Since its introduction in 2000, this approach has found application in many published studies.[21]

The $c(s)$ distribution is related to the well-known apparent sedimentation coefficient distribution $g^*(s)$ of nondiffusing particles, which was previously introduced in the form $g(s^*)$ arrived at through a data transformation dc/dt.[22,23] In comparison with the $g^*(s)$ distribution, it will be illustrated that the consideration of diffusion in the $c(s)$ distribution leads to a substantially increased resolution and provides a solution to the problem depicted in Fig. 1, the deconvolution of multiple sedimenting and diffusing protein species.

This chapter is intended as an overview of this approach in theory and practice. First, we describe the numerical methods for the continuous sedimentation coefficient distribution analysis. They are applied in the software SEDFIT (www.analyticalultracentrifugation.com) for analysis of experimental analytical ultracentrifugation data. Second, several variations are described that can give molar mass distributions, or permit the use of different prior knowledge to constrain the model, such as for proteins with

[19] P. Schuck, M. A. Perugini, N. R. Gonzales, G. J. Howlett, and D. Schubert, *Biophys. J.* **82**, 1096 (2002).

[20] P. Schuck, *Biophys. J.* **78**, 1606 (2000).

[21] http://www.analyticalultracentrifugation.com/references.htm.

[22] W. F. Stafford, *Methods Enzymol.* **240**, 478 (1994).

[23] J. S. Philo, *Anal. Biochem.* **279**, 151 (2000).

conformational changes or known molar mass, large particles, sedimenting cosolutes, compressible solvents, and others. Next, a brief description of the experimental procedures providing the best data basis for the $c(s)$ analysis is given followed by several examples of practical applications. Finally, its use for the study of interacting protein mixtures will be discussed.

Numerical Methods

Calculating sedimentation coefficient distributions from experimental data requires several numerical methods that address different aspects of sedimentation analysis. In the following, we will briefly review (1) the numerical solution of the Lamm equation for a single species, (2) how the model of the sedimentation boundary model can be adapted to the special noise structure of ultracentrifugation data, (3) the description of a continuous distribution by a Fredholm integral equation and different regularization methods, and (4) the approximation of the size-dependent diffusion coefficients.

Numerical Solutions of the Lamm Equation

One of the key components for the calculation of sedimentation coefficient distributions is the precise and efficient solution of the Lamm equation for a large range of s values. The evolution of the spatial distribution of a single sedimenting species $\chi(r, t)$ in a sector-shaped solution column of the centrifugal field can be described by the Lamm equation

$$\frac{\partial \chi}{\partial t} = \frac{1}{r}\frac{\partial}{\partial r}\left[rD\frac{\partial \chi}{\partial r} - s\omega^2 r^2 \chi\right] \tag{1}$$

where r denotes the distance from the center of rotation, ω the rotor angular velocity, and s and D the macromolecular sedimentation and diffusion coefficient, respectively.[7] s and D are related to the molar mass M by the Svedberg equation

$$D = \frac{sRT}{M(1 - \bar{\nu}\rho)} \tag{2}$$

(with the protein partial-specific volume $\bar{\nu}$, the solution density ρ, the gas constant R, and the absolute temperature T).[4] A method for estimating D as a function of s for an ensemble of protein species with different sizes is described later.

The Lamm equation (1) can be solved by a finite element method, which was first introduced to ultracentrifugal sedimentation by Claverie *et al.*[24] and recently generalized to permit greater efficiency and stability.[11] The basic strategy consists in the approximation of the concentration $\chi(r, t)$ as a superposition

$$\chi(r,t) \approx \sum_{k=1}^{N} \chi_k(t)P_k(r,t) \tag{3}$$

of N triangular elements P_k, which are defined as stepwise linear hat functions on a grid of radial points $r_i(t)$

$$P_k(r,t) = \begin{cases} (r - r_{k-1})/(r_k - r_{k-1}) & r_{k-1} \leq r \leq r_k \\ (r_{k+1} - r)/(r_{k+1} - r_k) & r_k < r \leq r_{k+1} \\ 0 & \text{else} \end{cases} \quad \text{for } k = 2, ..., N-1 \tag{4}$$

(details on the elements at the boundary of the solution column can be found in Schuck[11]) and thereby reducing the solution of Eq. (1) to the determination of the evolution of the coefficients $\chi_k(t)$. These can be found by multiplication of Eq. (1) with the elements P_k and integration in radial coordinates from meniscus m to bottom b of the solution column:

$$\int_m^b \frac{\partial \chi}{\partial t} P_k(r,t)rdr = \int_m^b \frac{\partial}{\partial r}\left[r\left(D\frac{\partial \chi}{\partial r} - s\omega^2 r\chi\right)\right]P_k(r,t)dr \tag{5}$$

Integration by parts of the right-hand side (rhs), taking advantage that the inner parentheses represent the total flux, which vanishes at both ends of the solution column, leads to

$$\int_m^b \frac{\partial \chi}{\partial t}P_k(r,t)rdr = s\omega^2 \int_m^b \chi \frac{\partial P_k(r,t)}{\partial r}r^2 dr - D\int_m^b \frac{\partial \chi}{\partial r}\frac{\partial P_k(r,t)}{\partial r}rdr \tag{6}$$

Insertion of Eq. (3) leads to

$$0 = \sum_j \frac{\partial \chi_j}{\partial t}\int_m^b P_j P_k rdr + \sum_j \chi_j \int_m^b \frac{\partial P_j}{\partial t}P_k rdr$$
$$- \omega^2 s \sum_j \chi_j \int_m^b P_j \frac{\partial P_k}{\partial r}r^2 dr + D\sum_j \chi_j \int_m^b \frac{\partial P_j}{\partial r}\frac{\partial P_k}{\partial r}rdr \tag{7}$$

In the original Claverie approach,[24] the hat functions P_k are constant in time, which reduces Eq. (7) to a linear equation system for the coefficients $\chi_k(t)$. With the abbreviations

[24] J.-M. Claverie, H. Dreux, and R. Cohen, *Biopolymers* **14,** 1685 (1975).

$$\mathbf{B}_{jk} = \int_m^b P_j P_k r\, dr, \qquad \mathbf{A}_{jk}^{(1)} = \int_m^b \frac{\partial P_j}{\partial r} \frac{\partial P_k}{\partial r} r\, dr, \qquad \mathbf{A}_{jk}^{(2)} = \int_m^b P_j \frac{\partial P_k}{\partial r} r^2\, dr$$

(8)

we can write Eq. (7) as a matrix-vector equation

$$\mathbf{B}\frac{\partial \vec{\chi}}{\partial t} = \left[\omega^2 s \mathbf{A}^{(2)} - D\mathbf{A}^{(1)}\right]\vec{\chi}$$

(9)

Although Eq. (9) can be integrated with a discrete time interval Δt to give the evolution of $\vec{\chi}$, it is advantageous to use a Crank–Nicholson scheme.[25,26] Here, the fluxes in the rhs are evaluated in the middle during the time step Δt, which leads to a higher order accuracy of the time step.[26] Substitution of $\vec{\chi}$ in the rhs by $[\vec{\chi}(t) + \vec{\chi}(t + \Delta t)]/2$ gives

$$\vec{\chi}(t + \Delta t) = \left[2\mathbf{B} - \Delta t\left(\omega^2 s \mathbf{A}^{(2)} - D\mathbf{A}^{(1)}\right)\right]^{-1}$$
$$\left[2\mathbf{B} + \Delta t\left(\omega^2 s \mathbf{A}^{(2)} - D\mathbf{A}^{(1)}\right)\right]\vec{\chi}(t)$$

(10)

which permits larger time steps without loss of precision.[9]

Unfortunately, the use of static elements P_k is numerically stable and efficient only if the sedimentation fluxes are small compared to the diffusion fluxes. For large particles with high s value and small diffusion coefficient, the algorithm produces oscillations at the end of the solution column and in the vicinity of the sedimentation boundary, which can be overcome only by very time-consuming use of a very fine discretization in space and time. Therefore, a numerical method was designed that is not only stable in the limit of $D = 0$, but increases in efficiency for smaller diffusion coefficients. This can be accomplished in a natural way by introducing hat functions P_k on a moving grid adapted to the sedimentation process, where each radial grid point (except the first and last) migrates as

$$r_k(t) = r_{k,0}\alpha(t - t_0) = r_{k,0} \exp\{s_G\omega^2(t - t_0)\} \qquad \text{for } k = 2, ..., N - 1$$
$$r_1(t) = m, \qquad r_N(t) = b$$

(11)

i.e., like a nondiffusing particle sedimenting in the gravitational field with the sedimentation coefficient s_G using the notation $\alpha(t)$ to abbreviate this translation process. With the choice of the starting grid

$$r_{k,0} = m(b/m)^{(k-3/2)/(N-1)} \qquad \text{for } k = 2, ..., N - 1$$

(12)

[25] J. Crank and P. Nicholson, *Proc. Camb. Philos. Soc.* **43**, 50 (1947).
[26] W. H. Press, S. A. Teukolsky, W. T. Vetterling, and B. P. Flarnery, "Numerical Recipes in C." Cambridge University Press, Cambridge, 1992.

the grid has the unique property that after a time interval of propagation

$$\Delta t_{\text{swap}} = \left[\omega^2 s_G (N-1)\right]^{-1} \ln(b/m) \tag{13}$$

it is mapped precisely onto the starting grid

$$r_k(t_0 + \Delta t_{\text{swap}}) = r_{k+1,0} \qquad \text{for } k = 2, ..., N-2 \tag{14}$$

and the propagation $\alpha(t)$ reduces to a simple renumbering of the indices of r_k, which is computationally trivial. When solving the Lamm equation with the moving grid, the hat functions are time dependent

$$\frac{\partial P_k}{\partial t} = \omega^2 s_G \times \begin{cases} -r/(r_k - r_{k-1}) & r_{k-1} \le r \le r_k \\ r/(r_{k+1} - r_k) & r_k \le r \le r_{k+1} \\ 0 & \text{else} \end{cases} \qquad k = 3, ..., N-2 \tag{15}$$

and the second term of Eq. (7) does not vanish. The abbreviation $\mathbf{A}_{jk}^{(3)} = \int_m^b (\partial P_j/\partial t) P_k r dr$ and the transformation of variables $\rho(r,t) = r/\alpha(t-t_0)$ leads to

$$\mathbf{B} \frac{\partial \vec{\chi}}{\partial t} = \left[\omega^2 (s\mathbf{A}^{(2)} - s_G \mathbf{A}^{(3)}) - \frac{D}{\alpha(t-t_0)^2} \mathbf{A}^{(1)}\right] \vec{\chi} \tag{16}$$

This is a form analogous to Eq. (9), but generalized to a sedimenting frame of reference. At finite diffusion, an additional term $\alpha(t)^{-2}$ corrects the diffusion for the stretching of the reference frame. Again, it can be very efficiently solved with the Crank–Nicholson scheme, as the matrices are independent of time, except for the corner 2×2 submatrices if time intervals other than Δt_{swap} are used.[11] As designed, the rhs is zero for $D = 0$ and $s = s_G$, and thus Eq. (16) is trivial to solve, and it remains numerically stable and efficient at low D. Finally, for the limiting case of very large particles with negligible diffusion, an analytical solution of the Lamm equation can be used[27,28]:

$$\chi(s,r,t) = c_0 e^{-2\omega^2 st} \times \begin{cases} 0 & \text{for } r < me^{\omega^2 st} \\ 1 & \text{else} \end{cases} \tag{17}$$

In combination, the static finite element approach by Claverie and the moving frame of reference solution, both in the Crank–Nicholson scheme, allow the efficient and precise calculation of the macromolecular concentration profiles in the centrifugal field for the complete spectrum of s values. Because of greater freedom in the choice of the time steps, the

[27] H. Fujita, "Mathematical Theory of Sedimentation Analysis." Academic Press, New York, 1962.
[28] P. Schuck and P. Rossmanith, *Biopolymers* **54**, 328 (2000).

former is more efficient for small particles with small s and high D, while the latter is better for simulating sedimentation of larger particles with high s and small D. In practice, when calculating the kernels for the sedimentation coefficient distribution, an empirical threshold can be used to select the best algorithm.

The numerical algorithms described previously can be adapted to account for unavoidable experimental complications that can cause measurable deviations from an "ideal" sedimentation process [Eq. (1)]. These include the acceleration phase of the rotor, which can be modeled by a time-dependent rotor speed in Eqs. (9) and (16).[29] It also includes the compressibility of the solvent, which even for aqueous buffers at high rotor speeds can produce density gradients that lead to small but significant retardation of the macromolecular migration.[16] Compressibility can be accounted for by locally varying sedimentation coefficients, again approximated as superposition of hat functions, and an analytical extension of Eq. (17) for compressible solvents has been derived.[16] The finite element method also allows us to account for the sedimentation of cosolutes, which change the local solvent density and viscosity and again lead to locally varying macromolecular sedimentation and diffusion coefficients.[15] The approach can also be used without further complications to describe flotation[29] and alternative experimental configurations, such as analytical zone centrifugation.[9,30]

Accounting for the Noise Structure of Sedimentation Velocity Data

Analytical ultracentrifugation data from sedimentation velocity experiments exhibit noise components that are not random. It is well-known that interference optical data can be dominated by these signal offsets, but they are usually significant also for absorbance data. The systematic signal offsets are caused by the spatial imperfections of the optical components, and by time-dependent vibrations and/or virtual 2π phase shifts of the interference patterns, respectively. Accordingly, the systematic noise can be decomposed into a time-invariant component $a_{TI}(r)$ and a radial-invariant component $a_{RI}(t)$

$$a(r,t) = S(\{p\}, r, t) + a_{TI}(r) + a_{RI}(t) + \varepsilon(r, t) \qquad (18)$$

with the experimental data and random noise denoted as $a(r, t)$, and $\varepsilon(r, t)$, respectively, and with $S(\{p\}, r_i, t_j)$ denoting any model for the sedimentation boundary, which may in general depend on a set of parameters $\{p\}$. Although other effects clearly exist, such as higher-order vibration modes,

[29] M. A. Perugini, P. Schuck, and G. J. Howlett, *Eur. J. Biochem.* **269,** 5939 (2002).
[30] J. Lebowitz, M. Teale, and P. Schuck, *Biochem. Soc. Trans.* **26,** 745 (1998).

these two orthogonal components can describe the systematic components with a precision usually better than the random noise of the data
acquisition. (An exception is shown in Fig. 2, where some residual
higher-order systematic contributions are visible.) They are also sufficient
to eliminate the need for an optical reference in absorbance optical transport experiments, for example, sedimentation velocity and analytical
electrophoresis.[31]

It is possible to account for the noise components $a_{TI}(r)$ and $a_{RI}(t)$ alge-
braically.[32] It is described here for the time-invariant baseline noise $a_{TI}(r)$,
but it can be combined with an analogous procedure for radial-invariant
noise $a_{RI}(t)$. If we abbreviate $a_{TI}(r_i)$ as b_i, we can minimize by least-squares

$$\operatorname*{Min}_{\{p\},\, b_i} \sum_{i,\, j} \left\{ a(r_i, t_j) - [b_i + S(\{p\}, r_i, t_j)] \right\}^2 \tag{19}$$

leading to

$$b_i(\{p\}) = \bar{a}_i = \bar{S}_i(\{p\}) \tag{20}$$

where $\bar{a}_i = (1/N_s) \sum_j a(r_i, t_j)$ (the "average scan") and $\bar{S}_i(\{p\}) = (1/N_s)$
$\sum_j S(\{p\}, r_i, t_j)$ (the "average boundary model") with the total number of
scans N_s. Insertion of this into the least-squares problem for the calculation
of the remaining parameters $\{p\}$ of the boundary model leads to

$$\operatorname*{Min}_{\{p\}} \sum_{i,\, j} \left\{ [a(r_i, t_j) - \bar{a}_i] - [S(\{p\}, r_i, t_j) - \bar{S}_i(\{p\})] \right\}^2 \tag{21}$$

This shows that the boundary parameters can be modeled relative to the
"average" radial profile as a reference. (It should be noted that this is different from the pairwise differencing, which is used in other approaches for
sedimentation analysis,[33] which leads to a stronger amplification of the
random noise.) It also shows that the best-fit systematic noise estimate is
easily calculated by least squares if a sedimentation model is available.
$a_{TI}(r)$ and $a_{RI}(t)$ can slightly correlate with sedimentation at very small
sedimentation coefficients. This correlation is equivalent to that introduced
by a time-difference analysis[19] and can be reduced by acquiring data for a
large time interval, leading to a large boundary displacement.[31] It can be
seen from Eqs. (18) and (21) that modeling the sedimentation data is in-
variant under a transformation $\tilde{a}(r, t) = a(r, t) - \tilde{a}_{TI}(r) - \tilde{a}_{RI}(t)$. Therefore,
the best-fit estimate of the systematic noise components can be subtracted

[31] S. R. Kar, J. S. Kinsbury, M. S. Lewis, T. M. Laue, and P. Schuck, *Anal. Biochem.* **285**, 135 (2000).
[32] P. Schuck and B. Demeler, *Biophys. J.* **76**, 2288 (1999).
[33] W. F. Stafford, *Methods Enzymol.* **323**, 302 (2000).

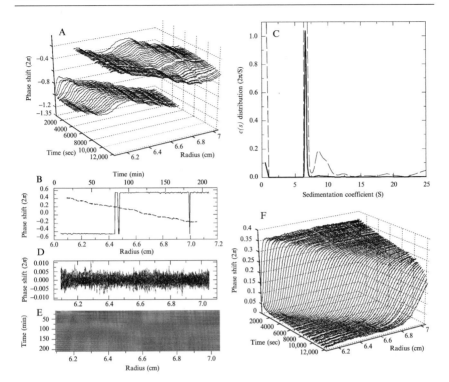

FIG. 2. Components of the $c(s)$ boundary model for experimental sedimentation velocity data. (A) The experimental raw data from laser interferometry optics (of an IgG sample dissolved in phosphate-buffered saline sedimenting at 40,000 rpm; only a data subset is shown). It is obvious that a substantial part of the signal is due to systematic noise contributions. The $c(s)$ boundary model accounts not only for the sedimenting macromolecules, but also for a time-invariant baseline component (TI noise) and an orthogonal radial-invariant baseline (RI noise) [Eq. (22)], which are simultaneously fit to the data. The results of this fit are shown in (B) to (F). (B) The best-fit TI noise as a function of radius (dashed line, lower abscissa) and the RI noise as a function of time (solid line, upper abscissa). Besides the apparent 2π phase shifts, the RI noise contains smaller amplitude vibrations. (C) The calculated $c(s)$ distribution (bold line, and in 20-fold magnification as dashed line). It has a sharp peak for the major monomeric IgG component, but also shows the presence of small but significant populations of higher oligomers and aggregates. (D, E) The residuals of the fit, which has an rms deviation of 0.0023 fringes. (D) The superimposed radial profiles for all times; (E) a bitmap representation of the residuals in the radius versus time plane (with a linear grayscale covering -0.02 to $+0.02$ from black to white). Vertical and horizontal structures are visible, which indicate the presence of small higher-order vibrations, but very little diagonal structure appears, which shows that there are no systematic residuals associated with the sedimenting boundary. (F) Since the data analysis is invariant under transformations that add systematic TI or RI noise, one can subtract the best-fit systematic noise components (B) from the raw data. This provides an equivalent of the raw data but free of systematic noise.

from the raw data without changing their information content [as long as systematic noise components according to Eq. (18) are still considered]. This can dramatically improve the possibility for visual inspection of the raw sedimentation data and the quality of their fit (Fig. 2).

The central observation exploited for calculating high-resolution sedimentation coefficients is that boundary spreading from differential migration and from diffusion is different, and that this difference can be extracted from the experimental data. To obtain reliable results, it is crucial to verify that the data are modeled within the random noise of the data acquisition. For testing the quality of a fit and if the residuals are within the statistical noise of the data acquisition, the overall rms error (and χ^2 statistics) can be used. The presence of remaining systematic deviations can be diagnosed with a runs test.[34] However, high Z values can be caused by both a systematic deviation of the model from the experimental data of the sedimentation boundary, or sometimes by optical imperfections, such as remaining higher-order vibrations in the data acquisition system. To help distinguish these cases, a bitmap representation was developed, in which the residuals encode the grayscale of pixels, which are placed in a picture according to the radius and time coordinate of each data point. Visual inspection then permits the detection of correlation of the residuals with the migration of the sedimentation boundary, as opposed to a static or periodic residuals pattern from optical imperfections. Although this is just an effective graphic representation of the residuals adapted to the data space of sedimentation velocity, it may be possible in future work to derive a quantitative measure for this correlation.

Distributions of Sedimentation Coefficients

A differential distribution of sedimentation coefficients $c(s)$ can be defined as the population of species with a sedimentation coefficient between s and $s + ds$. Accordingly, integration of the peaks of $c(s)$ can be used to calculate the weight-average s value s_w of the sedimenting components, and to obtain their partial loading concentrations. The distribution can be related to the experimentally measured evolution of the local concentration profiles throughout the centrifuge cell $a(r, t)$ by a Fredholm integral equation

$$a(r, t) = \int_{S_{min}}^{S_{max}} c(s)\chi(s, D(s), r, t)ds + a_{TI}(r) + a_{RI}(t) + \varepsilon \qquad (22)$$

[34] M. Straume and M. L. Johnson, *Methods Enzymol.* **210**, 87 (1992).

It states that the observed signal $a(r, t)$ is a simple linear superposition of the contributions of all subpopulations $c(s)$ at different s values (ranging from a minimal value $s_{\min}$ to a maximal value $s_{\max}$). Each species contributes to the radial and time-dependent signal with $\chi(s, D, r, t)$ as predicted by the Lamm equation for a macromolecule with sedimentation coefficient s and diffusion coefficient D. As will be outlined in detail later, we will estimate D through a functional dependence on s, assuming an average frictional ratio or hydrodynamic shape for all species. The noise contributions are described previously, and will be omitted for clarity in the following.

In principle, one could attempt to directly solve Eq. (22) by discretization of the sedimentation coefficients in a grid of N_s values s_k from $s_1, \ldots,$ s_N, and approximating $c(s)$ on this grid as a set of values $c(s_k)$ (short c_k, or in vector form $\vec{c}$). With data acquired at radius values r_i and at times t_j, Eq. (22) leads to the linear least-squares problem

$$a(r_i, t_j) \cong \sum_{k=1}^{N} c_k \chi(s_k, D_k, r_i, t_j) \tag{23}$$

Given the large number of data points, it is advantageous to solve Eq. (23) by normal equations with Cholesky decomposition.[35] Positivity of the c_k values can be ensured with the algorithm NNLS by Lawson and Hanson.[35] The drawback of this direct approach is that there may be in general many different distributions $c^*(s)$ that solve Eq. (22) nearly equally well and cannot be distinguished within the experimental precision of the data ε. This is a general, well-known problem of Fredholm integral equations,[36–38] and is true, in particular, for cases in which the kernel (the characteristic signal of a homogeneous subpopulation) is a smooth function.[37] Notoriously difficult are exponentials, such as the decay of the autocorrelation function in dynamic light scattering, for which much of the numerical strategies outlined in the following was originally introduced.[39] Fortunately, however, the large data space of sedimentation velocity and the characteristic features of the Lamm equation solutions $\chi(s, D, r, t)$ make the numerical determination of $c(s)$ less problematic.

To achieve a stable solution, regularization must generally be used, which allows exploiting additional prior knowledge or prior assumptions. Two different types of prior knowledge have been shown to be very

[35] C. L. Lawson and R. J. Hanson, "Solving Least Squares Problems." Prentice-Hall, Englewood Cliffs, NJ, 1974.
[36] D. L. Phillips, *Assoc. Comput. Mach.* **9,** 84 (1962).
[37] S. W. Provencher, *Comp. Phys. Commun.* **27,** 213 (1982).
[38] P. C. Hansen, *Inverse Probl.* **8,** 849 (1992).
[39] S. W. Provencher, *Makromol. Chem.* **180,** 201 (1979).

powerful in many other biophysical disciplines. Tikhonov–Phillips regularization (TP) selects from all possible solutions that fit the data well the one with the highest smoothness, calculated, for example, as $\int(dc/ds)^2 ds$. Alternatively, the maximum entropy (ME) principle selects the one that has the highest informational entropy or the minimal information content, given by $\int(c\log c)ds$.[40] For both, the rationale is that according to Occam's razor, the solution with the highest parsimony is to be preferred. ME can be justified by Bayesian principles, and will find a solution that is most likely, given a prior expectation (which in the absence of other data is commonly taken to be a uniform distribution).[26] Instead of solving the linear least-squares problem Eq. (23), we simultaneously minimize the residuals of the fit and optimize the parsimony constraint $H(\vec{c})$

$$\text{Min}_{c_k}\left\{\sum_{i,j}\left[a(r_i,t_j)-\sum_k c_k\chi(s_k,D_k,r_i,t_j)\right]^2+\alpha H(\vec{c})\right\} \quad (24)$$

where $H(\vec{c})=\sum_k c_k\log c_k$ for ME and $H(\vec{c})=\vec{c}^T\mathbf{D}^T\mathbf{D}\vec{c}$ (with D denoting a second derivative matrix) for TP regularization.[26] While the latter regularization method leads to a linear matrix equation, ME is a nonlinear problem that can be solved by the Levenberg–Marquardt method.[26]

One important question is, however, how to balance the parsimony constraint with the quality of the fit. This can be done following the approach first described by Provencher in the context of the program CONTIN.[41] The approach is based on the fact that all values of the regularization parameter $\alpha>0$ increase the rms error (or χ^2 value) of the fit relative to the (generally instable) least-squares optimum ($\alpha=0$). This allows use of a statistical criterion comparing the goodness of fit. The Fisher distribution predicts the probability of exceeding a ratio of $\chi^2(\alpha)/\chi^2(\alpha=0)$. Therefore, α can be iteratively adjusted such that the probability of $\chi^2(\alpha)/\chi^2(\alpha=0)$ corresponds to a predefined confidence level (usually 0.7). The degrees of freedom for the Fisher distribution are the number of data points (these are usually in the order of 10^5, and therefore the number of fitted parameters are negligible). Other methods for estimating the regularization parameter are possible.[42] The approach described ensures that the parsimony constraints are effective to suppress peaks in the $c(s)$ distribution that are not warranted by the data, but to extract all the details that are reliable.

[40] A. K. Livesey, P. Licinio, and M. Delaye, *J. Chem. Phys.* **84,** 5102 (1986).
[41] S. W. Provencher, *Comp. Phys. Commun.* **27,** 229 (1982).
[42] P. C. Hansen, *in* "Rank-Deficient and Discrete III-Posed Problems." SIAM, Philadelphia, PA, 1998.

The bias introduced by the regularization is generally small. However, it should be noted that the TP and ME methods can slightly differ in their properties and results. ME has the property that it can produce sharper peaks than TP and in our experience gives better results when dealing with a mixture of discrete protein species,[19] although we have observed that peaks in close vicinity sometimes appear to "attract." For broad and more continuous distributions, however, it is known to produce artificial oscillations, and TP appears to perform better in this case.[19]

It is clear that this analysis relies on the ability to model the experimental data well. In practice, this is usually the case, but some care may have to be taken with regard to the choice of the boundaries s_{min} and s_{max} of the distribution. If they are chosen too narrowly, they may constrain the model and produce unreliable results. Beyond the inspection of the fit and its residuals, this can be diagnosed by an increase of $c(s)$ toward the limits s_{min} or s_{max}. With regard to s_{max}, this situation can be indicative of the presence of large protein aggregates, and it is straightforward to increase the s_{max} value (to avoid high computational load when a large s range is spanned, the grid of s values can be chosen nonuniform, for example, logarithmic). At the lower limit, a value of $c(s_{min}) >> 0$ may be a result of either constraining s_{min} to a too high value, in which case it should be lowered, or a correlation with the baseline parameters. In our experience, the latter case is not detrimental to the reliability of the distribution at $s > s_{min}$, and therefore is preferred.

Estimating the Extent of Diffusion

So far, we have described the numerical methods for efficiently solving the Lamm equation, combining these solutions to calculate a sedimentation coefficient distribution, and adapting this boundary model to the special noise structure of sedimentation velocity data. The remaining problem, which was deferred previously, is to estimate the extent of diffusion for each species in the distribution Eq. (22). The best approach will depend on what is known about the ensemble of macromolecules under study. In principle, three different levels of complexity could be considered. The simplest approach would be to assume D to be constant for all species. This is a strong assumption, but it can be appropriate, for example, for distributions of proteins of the same Stokes radius, such as ferritin and apoferritin, and D could be taken from a measurement by dynamic light scattering. In the other extreme, one could consider the diffusion coefficients to be distributed themselves, independently of the sedimentation coefficients, and instead of a one-dimensional sedimentation coefficient distribution $c(s)$, a joint two-dimensional distribution $c(s, D)$ could be used to characterize the sample. This is the most general approach, but unfortunately a single sedimentation velocity experiment does not provide sufficient information. It is possible, however, to characterize such a $c(s, D)$ distribution by a global analysis of multiple sedimentation velocity experiments at different rotor speeds, or in global analysis with autocorrelation data from dynamic light scattering.[43]

An intermediate strategy is to estimate the diffusion coefficients from a monotonous single-valued function of the sedimentation coefficient. This can be expressed via the frictional ratio f/f_0, which is the ratio of the translational friction coefficient of the molecule relative to that of a sphere of the same mass and density. It is well-known that the values of f/f_0 are only very weakly dependent on the macromolecular shape[44] (examples for commonly observed values of the hydrated frictional ratio are 1.2–1.3 for relatively globular proteins, 1.5–1.8 for asymmetric or glycosylated proteins, and larger values for very asymmetric or unfolded proteins or linear chains). Therefore, for a given sample, it can be a good approximation to use as f/f_0 the weight-average frictional ratio $(f/f_0)_w$ of the macromolecules in the mixture. Using the Stokes–Einstein relationship and the Svedberg equation (2) we can derive

$$D(s) = \frac{\sqrt{2}}{18\pi} kTs^{-1/2}[\eta(f/f_0)_w]^{-3/2}[(1 - \bar{v}\rho)/\bar{v}]^{-1/2} \qquad (25)$$

(with k denoting the Boltzmann constant). This relationship can be used in Eq. (22) to calculate the Lamm equation solutions for each single species, followed by the determination of the best-fit sedimentation coefficient distribution $c(s)$. The correct value for $(f/f_0)_w$ can be determined iteratively by nonlinear regression, optimizing the quality of fit of the $c(s)$ boundary model as a function of the $(f/f_0)_w$ value. Except for experiments with unusually poor signal-to-noise ratio or shortened observation time, $(f/f_0)_w$ is well determined by the data. Very importantly, suboptimal values for the $(f/f_0)_w$ value are detrimental for the resolution of the $c(s)$ distribution, but have relatively little effect on the location of peaks in $c(s)$[19]: This is similar to the apparent sedimentation coefficient distribution $g^*(s)$ of nondiffusing particles, which corresponds to the limit $f/f_0 \to \infty$, and experiences (approximately Gaussian) broadening from the unaccounted macromolecular diffusion, but still reports the correct s value. To a much lesser extent, suboptimally corrected diffusion in the $c(s)$ distribution will cause a slight broadening of its peaks, but not a displacement.

[43] P. Schuck, in preparation (2004).
[44] C. R. Cantor and P. R. Schimmel, "Biophysical Chemistry. II. Techniques for the Study of Biological Structure and Function." W. H. Freeman & Co., New York, 1980.

In summary, we exploit three helpful properties for calculating the sedimentation coefficient distributions. First, the diffusion D depends only on the square root of s, which is a result of the sedimentation coefficient being much stronger size dependent than the diffusion coefficient. The dependence $D(s)$ can be expressed quantitatively through the frictional ratio f/f_0. Second, for mixtures of macromolecules of similar origin (i.e., folded proteins versus random polymer chains), f/f_0 also does not depend strongly on the detailed macromolecular shape. Third, if we approximate f/f_0 as a constant, the resulting $c(s)$ distribution is robust against small errors in $(f/f_0)_w$. As a result, we obtain a sedimentation coefficient distribution $c(s)$ that takes into account the diffusional spread for each species, and is therefore able to extract the heterogeneity and differential migration from the measured sedimentation boundaries.

Properties and Variations of the $c(s)$ Distribution

An important question is the sensitivity and precision of the $c(s)$ distribution, and the derived values for s_w and the partial loading concentrations. A safeguard against overinterpretation is the regularization, which provides only the simplest distribution that can model the data well. (It should be noted that this causes broadening of the peaks dependent on the noise and size of the raw data set; the peak width is therefore not always well suited as a characteristic of the sedimenting particles, and it is generally also not a good measure of the error in the s values.) Nevertheless, the distribution may contain a lot of details, since the data basis is very large, consisting in the order of 10^5 data points from the complete sedimentation process. Simulations show that from a set of profiles covering the complete sedimentation process at a signal-to-noise ratio of 200:1 (which can be readily achieved, e.g., at a loading concentration of 0.3–0.4 mg/ml of protein in the interference optics), minor peaks consisting of 0.2% of the total protein concentration can be reliably detected. The statistical accuracy of the calculated $c(s)$ distribution can be assessed by Monte Carlo simulations.[26] Assuming that the best-fit boundary model from the $c(s)$ fit is a good description of the data, and that the noise is normally distributed, a large number of synthetic data sets j can be generated and each subjected to the $c(s)$ analysis. The resulting family of $c_j(s)$ curves can be analyzed in different ways. First, for any s value s_k the limits of the central 68% of $c_j(s_k)$ values from the Monte Carlo simulation can be calculated. Applied to all s values, this procedure generates a one standard deviation contour for $c(s)$. For a detailed analysis with the intent to quantify the concentration and the s value of sedimenting components, an integral approach is advantageous. If each $c_j(s)$ curve is integrated from s_1 to s_2, the resulting statistics reflects the uncertainty of the partial concentration and s_w value of species sedimenting between s_1 and s_2.

Several variants of the $c(s)$ distribution have also proven useful in practice. For the study of the protein sedimentation coefficient distribution in the presence of a small-molecular-weight compound that contributes to the signal (e.g., a low concentration of nucleotides observed with the absorbance optics, or unmatched buffer salts in the refractometric optics), the sedimentation profiles can be characterized by a tilting baseline (from the approach of equilibrium for the small component) superimposed by the sedimenting boundaries of the protein components. In this case, the small component can frequently be described best as a species with discrete s and D values, superimposed to but not part of the $c(s)$ distribution [and thus not distorting $(f/f_0)_w$ and the regularization of the protein parameters]. Its sedimentation parameters can either be separately determined, or optimized in a nonlinear regression of the experimental data, jointly with the other non-linear parameters governing the $c(s)$ distribution, such as $(f/f_0)_w$ and the meniscus position of the solution column. Another variant of $c(s)$ is useful when a certain peak (between s' and s'') can be identified with a species of known molar mass: The subpopulation of species between s' and s'' can be excluded from the estimation $D(s)$ via Eq. (25), and instead the known molar mass can be used with the Svedberg equation to calculate D for these species. This additional constraint can be useful, for example, for the study of proteins with conformation changes.[45]

Since in the $c(s)$ distribution each s value is assigned an estimated D value, it is possible to transform $c(s)$ into a differential molar mass distribution $c(M)$. However, this requires caution; although the $c(s)$ distribution is not strongly influenced by the best-fit value for $(f/f_0)_w$, this is not true for $c(M)$. Here, a peak location will strongly depend on $(f/f_0)_w$ being a good estimate for that species. However, this can be fulfilled, for example, if it is known that the different species have the same hydrodynamic shape, or if the distribution consists of a single major peak [which will thus govern the weight-average $(f/f_0)_w$]. Despite this caveat, the $c(M)$ distribution can be a highly useful tool to obtain molar mass estimates of the sedimenting species frequently within 10% of the correct value.[46,47]

The $c(s)$ distribution can also be used to calculate apparent sedimentation coefficient distributions $g^*(s)$ of nondiffusing particles. This can be achieved either in the limit of a fixed very large value of $(f/f_0)_w$ (>10), or

[45] P. Schuck, Z. Taraporewala, P. McPhie, and J. T. Patton, *J. Biol. Chem.* **276,** 9679 (2000).
[46] D. M. Hatters, L. Wilson, B. W. Atcliffe, T. D. Mulhern, N. Guzzo-Pernell, and G. J. Howlett, *Biophys. J.* **81,** 371 (2001).
[47] J. Benach, Y.-T. Chou, J. J. Fak, A. Itkin, D. D. Nicolae, P. C. Smith, G. Wittrock, D. L. Floyd, C. M. Golsaz, L. M. Gierasch, and J. F. Hunt, *J. Biol. Chem.* **278,** 3628 (2003).

by substitution of the ordinary Lamm equation solutions with those for nondiffusing particles [Eq. (17)] or its extension to compressible solvents.[16] This $g^*(s)$ distribution has advantages over the distribution $g(s^*)$ derived by the *dcdt* method,[22,23,48] since the artificial broadening resulting from the approximation of dt by Δt is absent, permitting the analysis of larger data sets and larger boundary displacements between successive scans, such as in absorbance optics.[19,28]

In other variations, the $c(s)$ distribution was adapted to provide flotation coefficient distributions of spherical emulsion particles, making use of the known size dependence of the partial-specific volumes of the particles.[29] Also, the modeling of data from analytical zone centrifugation[30] by $c(s)$ has been implemented. In this configuration, a slightly higher correlation of the Lamm equation solutions was observed (in particular if the lamella thickness is treated as an unknown parameter), and stronger regularization appears to be required.

If the $c(s)$ analysis is applied to proteins that exhibit reversible interactions on the time scale of sedimentation, the $c(s)$ analysis does not reveal the populations of species sedimenting with a certain rate. This is because the fast reversible conversion of slower sedimenting and faster sedimenting protein complexes leads to an overall broadening of the sedimentation boundary, which can be empirically modeled only as an apparent superposition of sedimentation profiles of nonreacting species. In many cases, the chemical reaction results in artificial peaks in $c(s)$ at s values intermediate to those of the existing protein species, and in a bias of the $(f/f_0)_w$ value (usually toward unrealistically low values). Nevertheless, the $c(s)$ analysis can in this situation still reveal a large amount of important information. First, if experiments are conducted at different loading concentrations, a shift in the peak positions can reveal the presence of an interaction on the time scale of sedimentation. Second, the $c(s)$ distribution can allow us to distinguish the interacting components from those not participating in the reaction, such as small proteolytic degradation products, impurities, and/or irreversible protein aggregates, which can usually be well resolved from the native proteins and their reversible complexes.

Third, integration of $c(s)$ over the range of s values of the interacting species reveals a weight-average s value s_w, which is fully equivalent to that from second moment methods, and can be used for interacting systems to study the binding isotherm $s_w(c^*)$ of the interacting species.[17] [Interestingly, if the regularization is scaled as described previously and $c(s)$ provides a boundary model that is undistinguishable from that in the absence of regularization, it can be shown by second moment considerations that it

[48] W. F. Stafford, *Anal. Biochem.* **203**, 295 (1992).

does not affect the calculated s_w value, even though it may considerably change the $c(s)$ distribution itself.] Because of radial dilution, the effective concentration c^* for which the s_w value is determined is generally smaller than the loading concentration, but because the $c(s)$ distribution describes the entire sedimentation process, c^* is higher than the plateau concentration (for a precise analysis, see Schuck[17]). Because the $c(s)$ distribution can be based on a very large data base, good estimations of s_w can be accomplished at very low loading concentrations (with the loading signal only 2- to 3-fold the noise of the data acquisition), which can be crucial for the interpretation of the binding isotherm $s_w(c^*)$.[17]

Experimental Considerations

In contrast to the $g(s^*)$ distribution by $dcdt$[22,23,48] or the integral $G(s)$ distribution from the van Holde–Weischet method,[18,49,50] the $c(s)$ analysis does not require constraints in the data analysis to subsets of the sedimentation process. For $c(s)$, the resolution increases with increasing amount of data and observation time, and optimal data sets comprising the complete sedimentation process from directly after start of the centrifuge up to the depletion of the smallest visible species. This allows the characterization of species over a large range of s values in one experiment. No depletion at the meniscus and no established plateaus are required.

From theory, one could expect the presented approach to work best at high rotor speeds, where the resolution in s is highest, and the diffusional spread is smallest. In practice, however, the method also works well at lower rotor speeds, except for the approach to equilibrium of very heterogeneous samples (with species of different frictional ratios). To optimize the precision of the experimental data, a long temperature equilibration period ($\sim$1 h) of the resting rotor before start of the experiment is recommended, and an initial adjustment of the optics at a low rotor speed should be avoided. The acceleration of the rotor and the compressibility of the solvent can be accounted for when solving the Lamm equations. Because generally the meniscus position cannot be determined graphically with sufficient precision (it should be noted that it may not exactly coincide with the peak of the characteristic optical artifact), it is usually required to be included as a nonlinear fitting parameter that is optimized in a series of $c(s)$ analyses [jointly with $(f/f_0)_w$]. If the sedimentation data included regions of backdiffusion from the bottom of the solution column, or if a small molar mass species is included in the model, the bottom position of the solution column

[49] K. E. van Holde and W. O. Weischet, *Biopolymers* **17**, 1387 (1978).
[50] B. Demeler, H. Saber, and J. C. Hansen, *Biophys. J.* **72**, 397 (1997).

should also be determined by nonlinear regression. In most cases, a fit of the data within the noise of the data acquisition can be achieved.

Examples of Applications

The $c(s)$ method has found application in numerous studies of proteins and their interactions.[21] Here, we restrict the applications to only two examples. The first one will illustrate the use of $c(s)$ for an oligomeric protein that is stable on the time scale of sedimentation but exhibits impurities and microheterogeneity, and the second one will show the $c(s)$ analysis of a heterogeneous interaction between two proteins that form different complexes.

Figure 3 shows data from the study of the oligomeric state of the extracellular domain of a natural killer receptor, expressed by *Drosophila* cells. The molar mass calculated by amino acid composition is 43.7 kDa; however, the molar mass from MALDI was ~50.1 kDa, exhibiting several peaks ranging from 49.4 to 51.6 kDa as a result of differences in the extent of glycosylation. If the analysis is based on the assumption of the presence of a single sedimenting species, the spread of the sedimentation boundary can be approximately modeled as if arising from the single-species diffusion (Fig. 3A, dashed line). This leads to an s value of 4.83 S and a molar mass estimate of 63.4 kDa, suggesting that the protein would be predominantly monomeric. However, the measured s value exceeds the maximum possible s value (4.32 S) of a monomeric protein.

In contrast, the $c(s)$ analysis enables the decomposition of multiple sedimenting components and diffusional boundary spreading (Fig. 3C). The best-fit frictional ratio $(f/f_0)_w$ is 1.56, and the $c(s)$ distribution has a main peak at 4.74 S (Fig. 3E). Transformed into a $c(M)$ distribution, the main peak corresponds to a species of 94.3 kDa, close to that of a dimer. As discussed previously, although the determination of the molar mass is not generally possible, it provides a good estimate for the species sedimenting in a single main peak of the $c(s)$ distribution. The dimeric state of the protein was confirmed independently by sedimentation equilibrium. However, while the molar mass estimates from $c(s)$ are generally not as precise, more information is obtained here in comparison with sedimentation equilibrium: the $c(s)$ distribution also shows the presence of impurities, which include ~5% of species of higher molar mass (estimated 150–200 kDa), and ~8% of smaller species including a fraction in the size range of the monomer (possibly misfolded and incompetent monomer). Moreover, from the model $c(s)$ analysis, one can have access to the hydrodynamic shape of the molecules under study. For the given frictional ratio of 1.56, assuming hydration of 0.3 g/g, we can calculate axial ratios of 7.2 and 8.0 for the

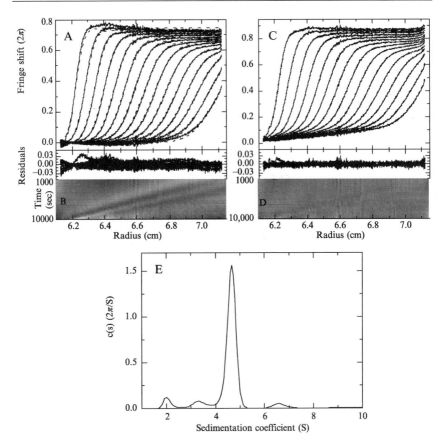

FIG. 3. Analysis of experimental sedimentation velocity data from the study of the oligomeric state of a glycosylated NK receptor fragment. Protein was dissolved in phosphate-buffered saline, and sedimentation profiles were observed at a rotor speed of 55,000 rpm and a rotor temperature of 22°. The radial protein distribution was observed with the interference optics in time intervals of 30 s. The partial-specific volume of the protein was estimated as 0.725 ml/g (based on the amino acid composition and the average extent of glycosylation as measured by MALDI). (A) Sedimentation profiles (every twentieth scan shown, solid lines) after subtraction of the systematic noise calculated with the best-fit single species Lamm equation model (dashed bold lines). (B) Superposition and bitmap representation of the residuals (rms error 0.0084 fringes). (C) Sedimentation profiles (solid lines) after subtraction of the $c(s)$ boundary model (dashed lines). (D) Superposition and bitmap representation of the residuals (rms error 0.0039 fringes). (E) $c(s)$ distribution.

hydrodynamically equivalent prolate and oblate ellipsoids, respectively. In the present example, however, a significant contribution to the hydrodynamic friction is due to the glycosylation, and the true molecular shapes can be expected to exhibit lower geometric asymmetry.

Noteworthy is the highly significant improvement of the quality of fit from an rms deviation of 0.0084 fringes and substantial systematic errors in the single species model (Fig. 3B) to 0.0039 fringes with very little systematic error (Fig. 3D) in the $c(s)$ analysis. This highlights the necessity to obtain a fit of high quality, and it demonstrates that the details in the experimental profiles contain the information necessary to distinguish boundary spreading due to diffusion from boundary spreading from diffusion combined with heterogeneity. In the present case, the heterogeneity arises from both the impurities and the microheterogeneity in the extent of glycosylation. Qualitatively similar results were obtained with protein expressed in *Drosophila* cells in the presence of tunicamycine, which suppresses the glycosylation. Like the fully glycosylated form, the $c(s)$ analysis revealed some inhomogeneity of the protein sample, but with a main peak at 83 kDa, close to 2-fold the monomer mass measured by MALDI (44.9 kDa). Here, the dimeric state was confirmed independently in combination with dynamic light scattering.

The second example illustrates the study of interacting species when the chemical reaction is at the time scale of sedimentation. The extracellular domain of a natural killer receptor (NKR) (15 kDa) is tested for binding *in vitro* with a class I MHC molecule (44 kDa).[51] These proteins were refolded from inclusion bodies expressed in *Escherichia coli* and are unglycosylated. Interestingly, no interaction was measurable with a surface plasmon resonance biosensor, likely due to immobilization artifacts. By sedimentation velocity and sedimentation equilibrium the NKR molecule is at micromolar concentrations a stable dimer [showing a single peak in $c(s)$ at 2.75 S; Fig. 3B]. At the same concentration, the MHC is monomeric with an s value of 3.66 S, but exhibiting some low molar mass contaminant at 1.75 S (Fig. 3B).

Figure 3A shows the sedimentation velocity data of a mixture of 43 μM NKR and 25 μM MHC (solid lines) and the best-fit $c(s)$ boundary model (dashed line). The corresponding $c(s)$ distribution has a large peak corresponding to the free NKR at 2.7 S, a smaller peak at 1.7 S corresponding to the contaminant of the MHC, and a bimodal peak in the range of 4–5 S (bold solid line in Fig. 4B) indicating complex formation and unambiguously demonstrating an interaction between the molecules. Although $c(s)$ appears to resolve two faster sedimenting components reflected in the bimodal nature of the complex peaks, it should be noted that when chemical reactions are observed on the same time scale as the sedimentation, the peaks in $c(s)$ do not necessarily reflect populations of species at the given s values (see previously). This is due to the coupled sedimentation of the

[51] J. Dam. *Nature Immunol.* **4,** 1213 (2003).

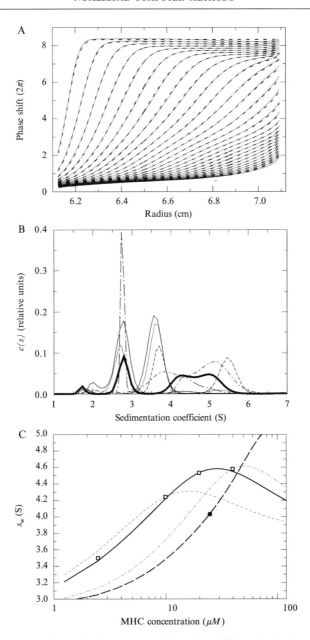

FIG. 4. Study of the interaction between an NKR and an MHC molecule by sedimentation velocity and $c(s)$ analysis. The experiments were conducted in phosphate-buffered saline at a rotor speed of 55,000 rpm, and a rotor temperature of $20\,^{\circ}$. (A) Sedimentation profiles measured with a mixture of 43 μM NKR and 25 μM MHC (solid lines, every tenth scan

reacting species, which leads to a broadening of the sedimentation boundaries in excess of their individual diffusional spread. In the $c(s)$ analysis, such a reactive boundary broadening results in a slight decrease in the quality of fit, and in a decrease of the best-fit $(f/f_0)_w$ value, which in the current example was 1.04, a value too small for a hydrated globular protein. A clear indication of the chemical reaction is the shift in the $c(s)$ distributions at different loading concentrations and molar ratios. Also shown in Fig. 4B are the results obtained from a series of experiments with a lower NKR concentration (10 μM fixed) and increasing concentrations of MHC (2.5, 10, 20, and 40 μM). While with excess NKR and MHC we find well-resolved peaks at the s values of the independently sedimenting molecules, the peaks of the complex are broad and shift from $\sim$4.3 S to 5.5 S.

The range of complex peaks at higher s values in the $c(s)$ distributions suggests the hypothesis of two different complex species with a 1:1 and a 2:1 stoichiometry of MHC per NKR dimer. However, it cannot be rigorously concluded at this stage of the analysis because the $c(s)$ peaks, when modeling a reaction boundary, do not correspond to s values of sedimenting species. Further, the largest observed s value of 5.5 S is smaller than the maximal s value of 6.3 S for a smooth and compact molecule of the molar mass corresponding to a 1:1 complex (MHC/NKR dimer). However, the available crystallographic structure corresponds to a complex of one NKR dimer symmetrically binding two MHC molecules. Although this structure was obtained by symmetry of one NKR monomer bound to one MHC molecule, the dimeric state of the NKR is unambiguous and therefore the 2:1 complex appears relevant. Hydrodynamic modeling of the 1:1 and 2:1 complexes based on the crystallographic structure[52] leads to predicted s values of 4.7 S and 6.3 S, respectively. On this basis, the largest observed s value of 5.5 S exceeds that of a 1:1 complex and appears to reflect

---

[52] J. Garcia De La Torre, M. L. Huertas, and B. Carrasco, *Biophys. J.* **78,** 719 (2000).

---

shown), and best-fit $c(s)$ boundary model (bold dashed line), which converges to a best-fit $(f/f_0)_w$ value of 1.04. Best-fit systematic noise components are subtracted for clarity. (B) normalized $c(s)$ distributions obtained from the data of the mixture shown in (A) (bold solid line), and from NKR and MHC alone (thin solid lines). Also shown are $c(s)$ distributions from a series of experiments with NKR at 10 μM and MHC at 2.5 μM (long dash-dotted line), 20 μM (short dash-dotted line), and 40 μM (short dashed line). (C) Dependence of the weight-average s value on MHC concentration with NKR at 43 μM (solid circle, single data point only) and at 10 μM (squares). The lines show the best-fit s_w isotherms for a 2:1 MHC per NKR dimer model (bold lines; solid line for NKR at 43 μM, dashed line for NKR at 10 μM), and for a 1:1 MHC per NKR dimer model (thin lines; short dash-dotted line for NKR at 43 μM, short dashed line for NKR at 10 μM). For more details on the experimental system and its analysis, see Dam.[51]

a mixture of populations of 1:1 and 2:1 complexes in a fast reversible inter-action. This illustrates the difficulty of interpreting the peak positions of $c(s)$ in the presence of interactions on the time scale of sedimentation, but also demonstrates how qualitative characteristics can be diagnosed from $c(s)$, in particular in conjunction with crystallographic data. (In the absence of reactive conversion of species during sedimentation, the $c(s)$ distribution will exhibit concentration independent peak positions.)

A rigorous thermodynamic analysis of the interaction is possible on the basis of the isotherms of weight-average sedimentation coefficients as a function of protein concentration.[2,17,53] These can be obtained from inte-grating the $c(s)$ distribution.[17] The integration can exclude the small con-tamination at 1.75 S, as this peak is clearly resolved and this species does not participate in the interaction. This emphasizes that the precise determination of s_w benefits from the high resolution of the $c(s)$ distribution. The s_w data as a function of total MHC concentration are shown in Fig. 4C (solid circle for NKR at 43 μM, and open squares for the data with NKR of 10 μM). Theoretical models for the s_w isotherms can be derived based on the laws of mass action and mass conservation, both for a single site model and for a model with two identical sites for MHC per NKR dimer. In con-trast to ordinary binding isotherms, the isotherms of weight-average s value typically exhibit a maximum at an optimal concentration and molar ratio, and decrease if either protein is present in excess concentration. These iso-therms were fitted globally to the experimental data, using the s values of the 1:1 and 2:1 complex (within constraints set by hydrodynamics) and the bind-ing constants as unknown parameters. While the single-site isotherms cannot describe the data well, the isotherms for a two-site model provide an excellent fit, with a best-fit K_d of 1.7 μM (0.2–1.7 μM) and 4-fold (2- to 57-fold) negative cooperativity for (Fig. 4C, bold solid and dashed line) and best-fit s values for the 1:1 and 2:1 complex of 4.30 S and 6.18 S, respect-ively. From these numbers, we find that the two binding sites of the NKR dimer are not saturated in our experimental conditions, considering that the highest s value observed was only 5.5 S. Although a higher saturation of the complex would be desirable in the analysis of the isotherm, nonideal sedimentation prevents the use of higher protein concentrations in the pre-sent case. Nevertheless, because the s_w isotherms of the two-site models ex-hibit broader peaks than is possible with single-site isotherms, the available data clearly demonstrate the existence of both 1:1 and 2:1 complexes in so-lution. Interestingly, although sedimentation equilibrium clearly showed the existence of a 1:1 complex, it was consistent with but did not permit the unambiguous detection of a 2:1 complex. More details on the analysis of this interacting protein system can be found in Dam.[51]

[53] J. J. Correia, *Methods Enzymol.* **321,** 81 (2000).

Discussion

In recent years, the development of fast finite element methods for solving the Lamm equation in combination with algebraic techniques for the detection and elimination of systematic noise in sedimentation velocity experiments enabled high-resolution, diffusion deconvoluted sedimentation coefficient distributions $c(s)$. They are solutions of Fredholm integral equations of the first kind, calculated with maximum entropy or Tikhonov–Phillips regularization, and based on approximations of the relationship between the diffusion and sedimentation coefficient $D(s)$ derived from hydrodynamic considerations. The current review provides a summary of the computational approaches, together with a discussion of experimental requirements and examples for the application to mixtures of noninteracting and interacting proteins.

In comparison with existing methods for calculating sedimentation coefficient distributions, perhaps the most basic difference is that $c(s)$ is a direct model for the complete data from the sedimentation process. In contrast, previous approaches include the transformation of time-derivative dc/dt to form an apparent sedimentation coefficient distribution $g(s*)^{22}$ (which does not distinguish diffusion and sedimentation), and the integral sedimentation coefficient distribution $G(s)$, which is obtained via an extrapolation procedure to infinite time.[49] Although the latter does provide a correction for diffusion, it can resolve only species that exhibit well-separated sedimentation boundaries. Both methods require specific experimental configurations and restricting the analysis to a subset of the available data. A detailed theoretical analysis of the relationships between $c(s)$, $g*(s)$, and $G(s)$ and a comparison of their results for different theoretical and experimental model systems can be found in Schuck $et\ al.$[19]

The use of numerical solutions of the Lamm equation in $c(s)$ permits a more flexible and general approach, analogous to continuous parameter distributions well-known in other biophysical techniques, such as dynamic light scattering[39,40] (implemented, for example, in the program CONTIN), fluorescence anisotropy,[54] and more recently optical affinity biosensing.[55] The common computational aspect of these continuous distributions is the description of the experimental data via a Fredholm integral equation, and the use of a Bayesian principle to calculate the most parsimonious distribution consistent with the data. In comparison with other techniques where the kernel (the characteristic function of a single species) is frequently exponential, the sedimentation coefficient distributions can

[54] P. J. Steinbach, $Biophys.\ J.$ **70,** 1521 (1996).

[55] J. Svitel, A. Balbo, R. A. Mariuzza, N. R. Gonzales, and P. Schuck, $Biophys.\ J.$ **84,** 4062 (2003).

achieve relative high resolution due to the very low correlation between different Lamm equation solutions. However, the $c(s)$ distribution does depend on additional global parameters [e.g., $(f/f_0)_w$ in the hydrodynamic approximation of $D(s)$], which require optimization by nonlinear regression. The latter step can be avoided in a more general approach with the model of a two-dimensional size-and-shape distribution [e.g., $c(s, D)$], but this requires global analysis and additional data from dynamic light scattering and/or sedimentation velocity experiments at different rotor speeds.[43]

The $c(s)$ method has already found applications in a variety of protein studies.[21] As illustrated, $c(s)$ can be used in the determination of the protein sedimentation coefficients and molar mass, and to detect trace components of protein aggregates and small and large molar mass impurities (Figs. 2 and 3), and a series of $c(s)$ distributions obtained at different protein concentrations can be used for the study of homogeneous and heterogeneous protein interactions (Fig. 4).[19]

Acknowledgments

We thank Drs. Roy Mariuzza and Klaus Karjalainen for their support and contributions to studies on the NK receptor systems.

[13] Analysis of Heterogeneous Interactions

By JAMES L. COLE

Introduction

Identification and characterization of critical macromolecular interactions within the cell are central research problems of the postgenomic era. High throughput mapping of protein–protein interactions is providing a global picture of the cellular interaction networks.[1] However, quantitative methods capable of accurately defining the stoichiometry, affinity, cooperativity, and thermodynamics are required for a fundamental mechanistic understanding and to effectively target molecular interactions for therapeutic intervention. Although many biologically significant interactions involve association of identical subunits (homogeneous interactions), a much larger class of binding events involves interactions of dissimilar partners (heterogeneous interactions or mixed interactions). Rigorous investigation

[1] C. von Mering, R. Krause, B. Snel, M. Cornell, S. G. Oliver, S. Fields, and P. Bork, *Nature* **417**, 399 (2002).

of heterogeneous interactions presents particular challenges for experimental design, data analysis, and interpretation. In this chapter, we present the mathematical formalism to describe heterogeneous equilibria, with particular attention to the relationship between the macroscopic and microscopic equilibrium constants. We then compare features of several biophysical methods that are commonly used to characterize heterogeneous interactions. Equilibrium and velocity analytical ultracentrifugation are particularly useful as baseline methods to define assembly models and extract equilibrium parameters. A chapter in this volume describes new methods for analysis of interacting systems by velocity sedimentation[2]; here, we focus on sedimentation equilibrium and illustrate this approach with an analysis of a nonspecific protein–nucleic acid interaction.

Formalism for the Analysis of Heterogeneous Interactions

Heterogeneous interactions involve at least two distinct macromolecular components. For simplicity, we consider systems composed of just two components, designated as A and B, that combine to form one or more species, designated A_iB_j. In some cases A or B self-associate as well, and either i or j can be zero. Note that each species A_iB_j may also exist in multiple conformational states with distinct hydrodynamic properties; here we consider only equilibria among species with differing composition.

The equilibria governing the association of A and B to form a family of complexes A_iB_j can be expressed as

$$i\text{A} + j\text{B} \overset{K_{ij}}{\leftrightarrow} A_iB_j \tag{1}$$

where the equilibrium constants, K_{ij}, are defined on the molar concentration scale according to the mass action law:

$$K_{ij} = \frac{[A_iB_j]}{[\text{A}]^i[\text{B}]^j} \tag{2}$$

Note that macromolecular association constants are often expressed in a weight/volume concentration scale in the analytical ultracentrifugation literature. Conversion among these different scales is straightforward.[3] It is important to realize that K_{ij} represents the *overall* equilibrium constant between the monomeric forms of A, B, and the complex A_iB_j. Often, assembly proceeds via sequential formation of a series of complexes of increasing

[2] J. Dam and P. Schuck, *Methods Enzymol.* **384**, 185 (2001).
[3] A. P. Minton, *Prog. Colloid Polym. Sci.* **107,** 11 (1997).

stoichiometry. For example, if a monomer of A contains multiple binding sites for interaction with B the following equilibria will pertain:

$$A + B \overset{K_1}{\leftrightarrow} AB, \qquad AB + B \overset{K_2}{\leftrightarrow} AB_2, \ldots, \qquad AB_{j-1} + B \overset{K_j}{\leftrightarrow} AB_j \qquad (3)$$

where $K_1, K_2, \ldots, K_j$ represent the *stepwise* equilibrium constants. The overall equilibrium constant K_{ij} is the product of each of the stepwise constants. As described below, analysis of these stepwise equilibrium constants can provide insight into the assembly mechanism.

Discrete Interactions

It has long been recognized in multisite ligand-binding systems that the experimentally accessible, macroscopic equilibrium constants are related to the intrinsic binding constants governing the microscopic equilibria by statistical factors.[4] These same considerations apply to more complex heterogeneous macromolecular interactions. The stepwise, macroscopic binding constant K_{ij} represents the product of an intrinsic binding constant k and a statistical factor related to the number of microscopic configurations of A, B, and the complex A_iB_j. For example, consider the $A + 2B \leftrightarrow AB_2$ system depicted in Fig. 1A, where A contains two identical binding sites for B. This model is commonly used to analyze bivalent antibody–antigen interactions.[5] There is only a single configuration for the free A and B species. However, in the AB complex there are two microscopic states with B bound to the left or right site, respectively. Thus, the AB state has a statistical weight of 2 and the stepwise, macroscopic binding constant $K_1 = 2k$. There is only a single configuration for the AB_2 complex and $K_2 = k/2$. Thus, the statistical effects cause a sequential decrease in the macroscopic binding constants with increasing saturation. In the general case where A contains s identical binding sites capable of binding B, the statistical factor C_x for the complex AB_x is given by[6]

$$C_x = \frac{s!}{(s-x)!x!} \qquad (4)$$

The overall, macroscopic binding constant defining the equilibrium is then given by

[4] J. T. Edsall and J. Wyman, "Biophysical Chemistry." Academic Press, New York, 1958.
[5] M. L. Doyle, M. Brigham-Burke, M. N. Blackburn, I. S. Brooks, T. M. Smith, R. Newman, M. Reff, W. F. Stafford, R. W. Sweet, A. Truneh, P. Hensley, and D. J. O'Shannessy, *Methods Enzymol.* **323**, 207 (2000).
[6] C. R. Cantor and P. R. Schimmel, *in* "Biophysical Chemistry" (C. R. Cantor and P. R. Schimmel, eds.), p. 849. W. H. Freeman and Co., New York, 1980.

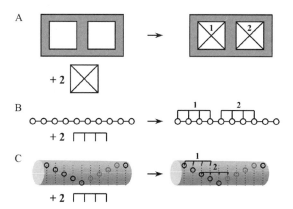

FIG. 1. Schematic illustration of specific and nonspecific heterointeractions. (A) Discrete binding with $A + 2B \rightarrow AB_2$ model. (B) Nonspecific binding of a large ligand to a finite, one-dimensional lattice. One of the six possible configurations for binding of two ligands of length 4 to a linear lattice of length 10. The open circles represent lattice sites. (C) Overlapping ligand model for the nonspecific binding of two ligands of length 4 to a helical lattice of length 10 with a minimal offset of 2. The leftmost binding site on the ligand must contact a lattice site indicated by an open circle. Only one of the 15 possible configurations is shown.

$$\prod_{j=1}^{x} K_j = C_x k^x \tag{5}$$

Or, in terms of the stepwise binding constants,

$$K_j = \frac{C_x}{C_{x-1}} k \tag{6}$$

Procedures for enumerating the statistical weights for more complex association models have been presented.[7]

When analyzing experimental data for multiple equilibria, it is useful to initially fit data in a model-independent fashion, where each of the equilibrium constants is treated as an independent parameter. The relative values of the equilibrium constants are then compared with those predicted by statistical considerations to test the association model. Deviations from these predictions are an indication that the model is incorrect. Possible causes are that the binding sites are not equivalent or interactions between sites (cooperativity). This approach may not be feasible for more complex models, where a large number of adjustable parameters gives rise to extensive cross-correlation among the equilibrium constants and unacceptably broad confidence intervals. In this case, it is more productive to directly

[7] M. L. Johnson and M. Straume, *Methods Enzymol.* **323**, 155 (2000).

fit for the intrinsic binding constants using the statistical factors to generate the macroscopic constants in the fitting function.

Nonspecific Interactions

In contrast to the discrete binding phenomena described previously, in some heterogeneous systems one component (A) consists of a polymeric lattice of identical or near-identical subunits and B interacts with one or more of these subunits without distinguishing among different positions in the lattice. For obvious reasons this type of binding is termed "nonspecific." Biologically significant examples include protein–nucleic acid interactions,[8] drug–DNA binding,[9] and protein binding to polysaccharide lattices.[10] Often the binding site for B consists of two or contiguous subunits. This situation is illustrated in Fig. 1B, where the open circles represent the repeating subunits in a polymeric molecule A and B occupies four contiguous lattice sites. The overlapping character of these sites results in the exclusion of binding of adjacent ligands, giving rise to binding isotherms that appear anticooperative. McGhee and von-Hippel derived the binding isotherm for this model in the case of an infinite, one-dimensional lattice[11]

$$\frac{v_{ns}}{K_{ns}[B]} = (1 - Nv_{ns})\left(\frac{1 - Nv_{ns}}{1 - (N-1)v_{ns}}\right)^{N-1} \tag{7}$$

where v_{ns} is the number of nonspecifically bound ligands (B) per lattice site, K_{ns} is the intrinsic binding constant for B interacting with an isolated region of A, [B] is the concentration of *free* B, and N is the number of lattice sites occluded by B.

Many binding experiments are performed using short lattices of defined length, such as oligonucleotides.[12–16] Here, a more relevant approach is to treat A as a finite one-dimensional lattice. An approximate isotherm for the finite lattice model has been presented[17,18]

[8] A. Revzin, "The Biology of Nonspecific DNA-Protein Interactions." CRC Press, Boca Raton, FL, 1990.

[9] J. J. Correia and J. B. Chaires, *Methods Enzymol.* **240,** 593 (1994).

[10] P. D. Munro, C. M. Jackson, and D. J. Winzor, *J. Theor. Biol.* **203,** 407 (2000).

[11] J. D. McGhee and P. H. von Hippel, *J. Mol. Biol.* **86,** 469 (1974).

[12] C. Schmedt, S. R. Green, L. Manche, D. R. Taylor, Y. Ma, and M. B. Mathews, *J. Mol. Biol.* **249,** 29 (1995).

[13] K. Wojtuszewski, M. E. Hawkins, J. L. Cole, and I. J. Mukerji, *Biochemistry* **40,** 2588 (2001).

[14] P. C. Bevilacqua and T. R. Cech, *Biochemistry* **35,** 9983 (1996).

[15] M. K. Levin and S. S. Patel, *J. Biol. Chem.* **277,** 29377 (2002).

[16] J. W. Ucci and J. L. Cole, *Biophys. Chem.,* in press (2004).

[17] I. R. Epstein, *Biophys. Chem.* **8,** 327 (1978).

[18] O. V. Tsodikov, J. A. Holbrook, I. A. Shkel, and M. T. Record, Jr., *Biophys. J.* **81,** 1960 (2001).

$$\frac{v_{ns}}{K_{ns}[B]} = (1 - Nv_{ns})\left(\frac{1 - Nv_{ns}}{1 - (N-1)v_{ns}}\right)^{N-1}\left(\frac{M - N + 1}{M}\right) \tag{8}$$

where M is the size of the lattice. Equation (8) is a good approximation to the exact solution over the entire saturation range even for lattices only slightly larger than the site size N.[18] A useful feature of this expression for fitting experimental data is that the site size does not have to be an integer and thus can be treated as an adjustable parameter in nonlinear least-squares analysis. This finite lattice model has also been extended to accommodate the combination of a specific binding site and nonspecific interactions.[18] One limitation of this approach is that Eq. (8) does not readily permit calculation of the statistical factors and populations of partially liganded species. In contrast, exact, combinatorial expressions have been derived to describe noncooperative and cooperative binding of large ligands to finite, one-dimensional lattice. The statistical weights for each ligation state are given by[10,17,19]

$$C_x = \frac{(M - Nx + x)!}{(M - Nx)!x!} \tag{9}$$

For example, in the case depicted in Fig. 1B with $M = 10$, $N = 4$, there are four ways to arrange two molecules of B on a lattice A and thus $C_2 = 4$. The advantage of using this formalism is that it allows one to calculate the statistical weight for each species in solution rather than simply the saturation of the lattice with component B. This approach is more appropriate for techniques, such as analytical ultracentrifugation, that are capable of resolving such partially liganded species. A restriction is that N cannot be treated as an adjustable parameter.

In recent studies of the nonspecific binding of protein kinase R (PKR) to double-stranded RNA sequences higher binding stoichiometries are found than can be explained within the one-dimensional finite lattice model.[16] Based on structural data, a new model was proposed in which the lattice supports overlapping ligand binding to different faces of the double helical RNA. Consecutive ligands initiate binding in fewer than N bases and thus the ligands overlap along the primary sequence of the lattice. This situation is depicted in Fig. 1C, which illustrates a configuration in which a ligand that occupies four lattice sites binds to a helical lattice of length 10. The open circles represent the leftmost edge of a binding site, e.g., the minor groove of a double-stranded nucleic acid. In this model, we define a minimum offset, Δ, which refers to the minimum number of lattice sites at which consecutive ligands can initiate binding ($\Delta = 2$ in Fig. 1C).

[19] S. A. Latt and H. A. Sober, *Biochemistry* **6**, 3293 (1967).

The value of Δ is determined by the size of the ligand and the geometry of the lattice. In this model, the limiting stoichiometry (s) is given by the largest integer value of S that satisfies the following inequality:

$$S \leq \frac{M - N}{\Delta} + 1 \tag{10}$$

Finally, the statistical weights are given by

$$C_x = \frac{[M - N - (x - 1)\Delta + x]!}{[M - N - (x - 1)\Delta]! x!} \tag{11}$$

The examples in Fig. 1 illustrate that the presence of ligand overlap significantly increases both the stoichiometry and statistical factors for nonspecific interactions.

Our discussion has emphasized the role of statistical factors in the analysis of apparent binding constants for both discrete and nonspecific heterointeractions. Obviously, these two classes of binding models differ substantially and interpretation of the relative values of the macroscopic constants should be done in the context of the appropriate model. Although PKR is known to bind to RNA nonspecifically, in some early studies binding constants were analyzed using statistical factors derived from the discrete model [Eq. (4)]; discrepancies between the observed data and the predicted ratios were interpreted as evidence for positive cooperativity.[12,14] In contrast, when PKR binding data are analyzed in the context of a more appropriate, nonspecific binding model there is no evidence for cooperativity.[16]

Experimental Methods

When beginning a project to characterize a molecular interaction one is faced with the problem of choosing an appropriate method. This can be a daunting task, since there are many experimental biophysical approaches available for quantifying heterogeneous macromolecular interactions. Here, we briefly consider the factors that govern the choice of methodology and consider some of the unique capabilities of each approach. At the outset, it should be acknowledged that no single approach is superior in all aspects, and as previously pointed out,[20] the most fruitful research strategies generally involve the use of a combination of methods.

Outside of the obvious considerations of the availability of instrumentation and personal experience, one of the key factors that govern the choice among alternative techniques is the type of information that is required by

[20] P. Hensley, *Structure* **4,** 367 (1996).

the researcher. Generally, simple evidence of a macromolecular interaction can be obtained using qualitative biochemical methods and the biophysical approaches are employed to obtain quantitative information, such as binding stoichiometries. At the next level of complexity, one may require equilibrium constants; in some cases, only a rank order of affinities is sufficient. Finally, it may be essential to obtain additional data such as the thermodynamic parameters (ΔH, ΔS, ΔC_p) and association and dissociation rates for the interaction.

The second set of considerations relates to sample-specific issues. First, how much material is available and how many samples must be assayed? Second, what are the properties of the material? The chief considerations are solubility and stability. Other issues relate to modification of the sample for analysis. Can the reagent be readily labeled with extrinsic probes for fluorescence measurements or immobilized for surface plasmon resonance? Do these modifications alter the reactivity of the sample? Finally, if one wishes to measure affinities, what is the expected range of K_d? This will determine sample concentrations and the experimental sensitivity that are required.

Given these considerations, Table I compares the information content, sensitivity, and unique features for several of the most popular biophysical methods. Admittedly, this comparison is somewhat superficial in that each of these approaches actually constitutes a diverse family of experimental configurations with differing detection methodologies, sensitivities, and sample requirements. In particular, we have combined together static and dynamic light-scattering measurements as well as the range of fluorescence-based methods (anisotropy, correlation spectroscopy, energy transfer). However, some interesting categories that differentiate these approaches do emerge and are summarized in the following.

1. Component properties. Several methods (sedimentation, light scattering, fluorescence) provide potentially useful information about the separate components (A and B) as well as the complexes.

2. Complex properties. It may be important to determine the absolute stoichiometry of a complex rather than a molar ratio of components; for example, it may be necessary to distinguish AB from A_2B_2. Techniques that are based on the mass or hydrodynamic properties of the complex provide this information.

3. Species resolution. In some methods, the experimental observable is the fractional saturation of A with B; in others, the separate species are resolved. A particularly attractive feature of sedimentation velocity is that multiple species are resolved based on differences in sedimentation coefficients and diffusion constants. Note that sedimentation velocity

TABLE I
BIOPHYSICAL METHODS TO CHARACTERIZE MACROMOLECULAR INTERACTIONS

	Sedimentation equilibrium[a]	Sedimentation velocity[b]	Isothermal titration calorimetry[c]	Surface plasmon resonance[d]	Light scattering[e]	Fluorescence[f]
Component properties	Mass	Mass and hydrodynamics	—	—	Mass and shape	Hydrodynamics
Complex properties	Stoichiometry	Stoichiometry	Molar ratio	Molar ratio	Stoichiometry	Molar ratio
Species resolution	Resolved complexes	Resolved complexes	Fractional saturation	Fractional saturation	Resolved complexes[h]	Fractional saturation
K_d range (M)	10^{-3}–10^{-9}	10^{-4}–10^{-8}	10^{-3}–10^{-8} (10^{-2}–10^{-12})[g]	10^{-3}–10^{-12}	—	10^{-3}–10^{-12}
Additional information	—	Kinetics and hydrodynamics	Thermodynamics	Kinetics	—	—
Material requirements	20–120 μl	400 μl	1.5 ml	μl	—	μl–nl

[a] References 3, 20, 21–26.
[b] References 2, 23, 26, 27, 29.
[c] References 30, 31.
[d] References 32, 33.
[e] References 34–36.
[f] References 37–39.
[g] Ligand displacement and proton linkage techniques greatly extend the accessible K_d range for low-affinity[40] and high-affinity[41,42] interactions measured by isothermal titration calorimetry.
[h] In dynamic light scattering resolution is achieved by analysis of the autocorrelation function as a continuous distribution of diffusing species.[43]

analysis methods based on weight-average sedimentation coefficients,[27,28] sedimentation coefficient distribution analysis,[2] and direct boundary analysis[29] are expanding the power of this method to characterize complex heterogeneous interactions. In sedimentation equilibrium, the species are resolved based on mass and optical properties and in dynamic light scattering analysis of the autocorrelation function yields a diffusion constant distribution function. However, both of these methods involve analysis of data using model functions consisting of sums of exponential terms, which can be an ill-conditional fitting problem.[44]

In summary, there are a variety of approaches to define heterogeneous interactions with complementary capabilities. In our opinion, analytical ultracentrifugation methods (equilibrium and velocity) fill a unique niche in the repertoire of techniques used to measure macromolecular interactions due to their rigor and ability to discriminate among alternative association models. Other methods are more sensitive and rapid. In particular, surface plasmon resonance and fluorescence approaches are easily configured for screening large numbers of samples. The more fundamental methods are most profitably used at the outset of a project to define the correct reaction scheme or model for the interaction and to accurately measure equilibrium

[21] J. L. Cole and J. C. Hansen, *J. Biomol. Tech.* **10**, 163 (1999).

[22] P. Schuck and E. Braswell, *Curr. Protocols Immunol.* Sect. 18.8.1 (2002).

[23] J. Lebowitz, M. S. Lewis, and P. Schuck, *Protein Sci.* **11**, 2067 (2002).

[24] T. M. Laue and W. F. Stafford, *Annu. Rev. Biophys. Biomol. Struct.* **28**, 75 (1999).

[25] J. S. Philo, *Methods Enzymol.* **321**, 100 (2000).

[26] G. Rivas, W. Stafford, and A. P. Minton, *Methods* **19**, 194 (1999).

[27] J. J. Correia, *Methods Enzymol.* **321**, 81 (2000).

[28] P. Schuck, *Anal. Biochem.* **320**, 104 (2003).

[29] W. F. Stafford and P. J. Sherwood, *Biophys. Chem.*, in press (2004).

[30] H. F. Fisher and N. Singh, *Methods Enzymol.* **259**, 194 (1995).

[31] P. C. Weber and F. R. Salemme, *Curr. Opin. Struct. Biol.* **13**, 115 (2003).

[32] D. G. Myszka, *Methods Enzymol.* **323**, 325 (2000).

[33] P. Schuck, *Annu. Rev. Biophys. Biomol. Struct.* **26**, 541 (1997).

[34] S. E. Harding, D. B. Sattelle, and V. A. Bloomfield, "Laser Light Scattering in Biochemistry." Royal Society of Chemistry, Cambridge. CRC Press, Boca Raton, FL, 1992.

[35] S. E. Harding, *Methods Mol. Biol.* **22**, 85 (1994).

[36] J. Wen, T. Arakawa, and J. S. Philo, *Anal. Biochem.* **240**, 155 (1996).

[37] D. M. Jameson and W. H. Sawyer, *Methods Enzymol.* **246**, 283 (1995).

[38] N. L. Thompson, A. M. Lieto, and N. W. Allen, *Curr. Opin. Struct. Biol.* **12**, 634 (2002).

[39] J. R. Lundblad, M. Laurance, and R. H. Goodman, *Mol. Endocrinol.* **10**, 607 (1996).

[40] Y. L. Zhang and Z. Y. Zhang, *Anal. Biochem.* **261**, 139 (1998).

[41] M. L. Doyle, G. Louie, P. R. Dal Monte, and T. D. Sokoloski, *Methods Enzymol.* **259**, 183 (1995).

[42] B. W. Sigurskjold, *Anal. Biochem.* **277**, 260 (2000).

[43] S. W. Provencher, *Comp. Phys. Commun.* **27**, 229 (1982).

[44] K. Holmstrom and J. Petersson, *Appl. Math. Comput.* **126**, 31 (2002).

constants. Subsequently, methods that are more rapid may be used to process a large number of samples. This paradigm has been employed in studies of the activation of the 2′,5′-oligoadenylate (2,5A)-dependent ribonuclease. Sedimentation equilibrium measurements defined an activation model in which one 2,5A activator (A) binds to a ribonuclease monomer (B) to produce the AB complex, which subsequently dimerizes to the active form, A_2B_2.[45,46] Subsequently, fluorescence polarization[45] and enzymatic activity assays[47] were used to obtain structure–activity relationships for large numbers of enzymatic activators.

Sedimentation Equilibrium

Sedimentation equilibrium is a rigorous and well-established method to define the stoichiometry and affinity of macromolecular interactions, and the basic principles have been the subject of several recent reviews.[21–24] Here, we consider the mathematical formalism for analysis of heterointeracting systems by sedimentation equilibrium, describe some practical issues for experimental design and computer fitting of experimental data using global nonlinear least squares, and present an example of a protein–RNA interaction. Other chapters in this volume describe new methods for analysis of interacting systems by velocity sedimentation.[2]

Analysis of heterogeneous interactions is considerably more difficult than self-association because the fitting models give rise to a larger number of adjustable parameters. The analyses are plagued by multiple minima and by unacceptably broad confidence intervals for the deduced parameters due to extensive cross-correlation. A variety of methods have been described to circumvent these problems.[3,25] In the case of protein–nucleic acid interactions, where the two reactants have markedly different absorption spectra, collection of radial absorption gradients at multiple wavelengths is particularly useful to accurately define the concentration of each of the components and to enhance sensitivity. Thus, there has been renewed interest in the use of sedimentation equilibrium in the quantitative analysis of protein–nucleic acid interactions.[13,16,45,48–51]

[45] J. L. Cole, S. S. Carroll, E. S. Blue, T. Viscount, and L. C. Kuo, *J. Biol. Chem.* **272,** 19187 (1997).
[46] J. L. Cole, S. S. Carroll, and L. C. Kuo, *J. Biol. Chem.* **271,** 3979 (1996).
[47] S. S. Carroll, J. L. Cole, T. Viscount, J. Geib, J. Gehman, and L. C. Kuo, *J. Biol. Chem.* **272,** 19193 (1997).
[48] M. F. Bailey, B. E. Davidson, A. P. Minton, W. H. Sawyer, and G. J. Howlett, *J. Mol. Biol.* **263,** 671 (1996).
[49] S. J. Kim, T. Tsukiyama, M. S. Lewis, and C. Wu, *Protein Sci.* **3,** 1040 (1994).
[50] M. S. Lewis, R. I. Shrager, and S.-J. Kim, *in* "Modern Analytical Ultracentrifugation" (T. M. Shuster and T. M. Laue, eds.), p. 94. Birkhauser, Boston, 1994.

Our preferred approach is to directly analyze sedimentation equilibrium concentration gradients by nonlinear least-squares fitting methods. Alternatively, secondary parameters, such as weight-average molecular weights or the omega function, can be extracted from the raw data and these secondary parameters can be fit to various association models.[3,52,53] Also, Lewis and co-workers have described a matrix method in which sedimentation equilibrium profiles at multiple wavelengths are used to construct molar concentration distributions of each reactant.[50] These concentration distributions are then jointly fit to an association model to obtain equilibrium constants. We prefer the direct fitting approach because it requires less manipulation of the data and the experimental uncertainty is statistically well defined.

For a single ideal species (A), the radial absorption gradient at sedimentation equilibrium is given by

$$A(r, \lambda) = \delta_\lambda + \varepsilon_{A,\lambda} C_{0,A} \exp[M_A^* \phi (r^2 - r_0^2)] \tag{12}$$

where $A(r,\lambda)$ is the radial-dependent absorbance at wavelength λ, δ_λ is a baseline offset, $\varepsilon_{A,\lambda}$ is the molar extinction coefficient of A, $C_{0,A}$ is the molar concentration of A at the arbitrary reference distance r_0, and M_A^* is the buoyant molecular weight of A. M_A^* is defined by

$$M_A^* = M_A(1 - \bar{v}_A \rho) \tag{13}$$

where M_A is the mass of A, $\bar{v}_A$ is the partial specific volume of A, and ρ is the solvent density. The factor Φ is given by

$$\phi = \frac{\omega^2}{2RT} \tag{14}$$

where ω is the angular velocity of the rotor in radians/s, R is the molar gas constant, and T is the absolute temperature. The baseline offset term arises from absorption mismatches between the sample and reference sectors due to the presence of nonsedimenting, absorbing contaminants or unequal oxidation of reductants such as DTT. The concentrations of each of the complexes A_iB_j are defined in terms of equilibrium constants and the concentrations of A and B using Eq. (2). The total radial concentration gradient $A(r,\lambda)$ is then written as the sum of the contributions from the free A, B, and each of the complexes:

[51] T. M. Laue, D. F. Senear, S. Eaton, and J. B. Ross, *Biochemistry* **32,** 2469 (1993).
[52] B. K. Milthorpe, P. D. Jeffrey, and L. W. Nichol, *Biophys. Chem.* **3,** 169 (1975).
[53] D. J. Winzor, M. P. Jacobsen, and P. R. Wills, *Biochemistry* **37,** 2226 (1998).

$$A(r, \lambda) = \delta_\lambda + \varepsilon_{A,\lambda} C_{0,A} \exp[M_A^* \phi(r^2 - r_0^2)] + \varepsilon_{B,\lambda} C_{0,B} \exp[M_B^* \phi(r^2 - r_0^2)]$$
$$+ \sum_{i,j} (i\varepsilon_{A,\lambda} + j\varepsilon_{P,\lambda}) C_{0,A}^i C_{0,B}^j \exp[(iM_A^* + jM_B^*)\phi(r^2 - r_0^2) + \ln K_{ij}]$$

$$(15)$$

Here, the K_{ij} terms are written as overall equilibrium constants and are expressed in the form exp [ln K_j] to constrain them to be positive. It is worth noting a few of the implicit assumptions contained within Eq. (15):

1. Reversible equilibrium. It is assumed that all of the species present in the sample participate in a reversible mass-action equilibrium. If there are monomeric or oligomeric species that are not in equilibrium or significant impurities present, additional terms must be added to accommodate this additional heterogeneity.
2. Absence of absorption changes. We assume that the molar extinction coefficient of the species A_iB_j is the composition-weighted sum of the extinction of the individual components A and B. The extent of hypo- or hyperchromism effects can easily be tested beforehand using absorption mixing experiments.
3. Absence of volume changes. In the absence of volume changes the buoyant molecular weight of the complex is the composition-weighted sum of the components: $iM_A^* + jM_B^*$.

Analysis of Sedimentation Equilibrium Data

When using Eq. (15) to fit experimental data it is critical to reduce the number of adjustable parameters to obtain reliable, unique, and well-defined estimates of the equilibrium constants. Typically, we measure the buoyant molecular weights of A and B in a series of independent experiments and fix these values for the analysis of the heterointeraction. This also serves as a quality-control step to ensure that the individual reactants are homogeneous and well behaved in solution. Sedimentation velocity experiments are also useful to determine homogeneity. In addition, where possible the baseline offsets are measured by overspeeding following the experiment. This is accomplished by increasing the rotor speed to 45,000 rpm following the experiment, waiting 4–10 h to allow the meniscus region to become depleted of macromolecular species, and then recording the absorption near the meniscus at each of the wavelengths of interest. It is not possible to completely deplete the meniscus of A or B when these components are lower molecular weight. In this case, estimates of the offsets can be obtained by fitting the radial absorption gradient near the meniscus using a single ideal species model [Eq. (12)], but this approach is less reliable.

It is critical that the relative molar extinction coefficients are accurately determined at each wavelength when performing global analysis on data obtained at multiple wavelengths. The accuracy of wavelength selection for the monochromator in the XL-A centrifuge is on the order of ± 2 nm, which can result in substantial errors, particularly at 230 nm, which is on a steeply rising shoulder of the protein absorbance. In each experiment we include at least two sample channels containing pure components, A and B, respectively. Once equilibrium is achieved, data are collected at each wavelength of interest. It is critical to scan all the cells at one wavelength before collecting data at the next wavelength; otherwise, the wavelength setting will not be consistent between the different cells. We typically calculate the absolute extinction coefficients for protein at 280 nm and nucleic acid at 260 nm based on composition or other experimental measurements. Then, the relative extinction coefficients at other wavelengths is experimentally determined in the XL-A centrifuge by jointly fitting absorption gradients obtained from channels containing either pure protein or nucleic acid using a procedure described by Lewis et al.[50] In this manner, a self-consistent set of extinction coefficients is obtained for each experiment.

There are several software options available for global analysis of heterogeneous interactions by sedimentation equilibrium. First, it should be noted that the popular WinNonlin software[54] (www.ucc.uconn.edu/~wwwbiotc/UAF.html) is limited to self-association models as are the ORIGIN-based software supplied with the Beckman-Coulter instrument and the ULTRASCAN package (www.ultrascan.uthscsa.edu). Two freely distributed packages for analysis of heterointeractions are TWOCOMP[3] (www.bbri.org/rasmb/rasmb.html) and ULTRASPIN (www.mrccpe.cam.ac.uk/ultraspin_intro.php). Although these packages are very useful for certain applications they are somewhat limited in the models that can be analyzed and the quantity of the data that can be simultaneously fit. Therefore, many researchers, including us, employ commercial mathematical software packages such as IGOR Pro (Wavemetrics, www.wavemetrics.com), MLAB (Civilized software; www.civilized.com), and MATLAB (Mathworks, www.mathworks.com). For complex association models, it can be useful to perform experimental simulations to test whether the parameters of interest are experimentally accessible and to ensure that the experimental conditions (loading concentrations, rotor speeds, wavelengths) are devised to maximize the information content. In particular, if the user is interested in measuring equilibrium constants with accuracy it is

[54] M. L. Johnson, J. J. Correia, D. A. Yphantis, and H. R. Halvorson, *Biophys. J.* **36,** 575 (1981).

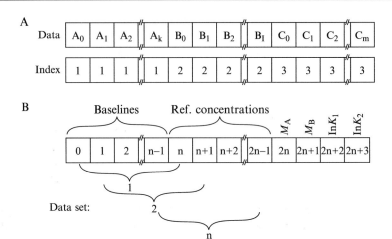

Fig. 2. Array designs used in global nonlinear least-squares fitting of sedimentation equilibrium experiment. (A) Design of the concatenated data arrays. Individual data sets are denoted A, B, C, with elements $0–k$, $0–i$, and $0–m$, respectively. The index vector labels each point in the array with the source data file. (B) Design of the parameter array for an $A + 2B \rightarrow AB_2$ model with n data sets.

important that the loading concentrations are chosen such that each of the species participating in the equilibria is significantly populated. A useful simulation package is available within the TWOCOMP software.

The general issues involved in global nonlinear least-squares analysis have been described in this series.[55] However, there are a few particular considerations for global analysis of heterogeneous interactions by sedimentation equilibrium: some parameters must be defined to be local to each centrifuge cell or radial absorbance scan and others must be global to the entire data set. Our implementation of global analysis (XLAnalysis) is a collection of functions and procedures within the IGOR Pro software. It uses a simple indexing scheme.[55] Each data file contains two arrays: the independent variable x (radius) and the dependent variable y (absorbance or fringe displacement). For global analysis the data sets from each of the channels, A,B,C, ... ,n, are concatenated to produce a single x and a single y array (Fig. 2). An integer index array is also created that labels each point in the concatenated array with the source data file. Additionally, arrays are also created that contain ancillary data [reference distances, rotor speeds, extinction coefficient(s)]. In the fitting procedure, certain parameters (baseline offsets, reference concentrations) are usually local to

[55] J. M. Beecham, *Methods Enzymol.* **210**, 37 (1992).

each channel and others (buoyant masses, equilibrium constants, stoichiometries) are global. In multiwavelength experiments, multiple radial absorption gradients may originate from the same physical sample cell and the reference concentrations for A and B that refer to the same sample should be constrained to be equal. The reference radii chosen for these data channels must be equal as well. The fitting parameters are passed to the fitting functions in an array that allows the function to discriminate between these local and global parameters using the index array and a global variable containing the number of fit files, n (Fig. 2). The array is designed such that it is independent of the number of files being fit but it is dependent on the nature of the fitting model.

There are several numerical algorithms commonly in use for parameter estimation by nonlinear least-squares analysis.[56-58] The original *Nonlin* program[54] utilizes a variation of the Gauss–Newton method that does not make the assumption of orthogonality at the initial stages of convergence.[57] The Marquardt algorithm is the most commonly used method and combines the steepest descent and Gauss–Newton methods.[56-59] In our experience, the two methods work equally well for systems in which a well-defined minimum is present in the error surface. Finally, the Simplex directed search method is less efficient but is guaranteed to converge.[60]

In many commercial software packages, the usual method to obtain parameter confidence intervals is based on the variance/covariance matrix, which can seriously underestimate the actual confidence intervals for nonlinear models. We use the F statistic as a more rigorous measure of significance of the increase in the value of the least-squares norm.[56-58] The confidence intervals for a parameter of interest are obtained by fixing it at some value slightly removed from the best fit and performing a fit where all of the other parameters are free to adjust to their best-fit values. The variance of this fit is calculated and the parameter is incremented or decremented until the ratio of the variance for this constrained fit to the variance for the best-fit value $(s^2/s^2_{minimum})$ equals the target ratio, defined by

$$\frac{s^2}{s^2_{minimum}} = 1 + \frac{M}{N - M} F(M, N - M, 1 - P) \tag{16}$$

[56] M. L. Johnson and M. Straume, *in* "Modern Analytical Ultacentrifugation" (T. M. Shuster and T. M. Laue, eds.), p. 37. Birkhauser, Boston, 1994.

[57] M. L. Johnson and L. M. Faunt, *Methods Enzymol.* **210,** 1 (1992).

[58] P. R. Bevington and D. K. Robinson, "Data Reduction and Error Analysis for the Physical Sciences." McGraw-Hill, New York, 1992.

[59] D. W. Marquardt, *SIAM J. Appl. Math.* **14,** 1116 (1963).

[60] J. A. Neider and R. Mead, *Comput. J.* **7,** 308 (1965).

where M is the number of parameters, N is the number of data points, F is the F statistic, and P is the probability that two fits are equivalent. The search is conducted for each parameter in turn using a bisection algorithm. This method makes no assumptions regarding the shape of the error surface and accounts for the cross-correlation (nonorthogonality) of the fitting parameters and the nonlinearity of the fitting equation. It is also helpful to visualize the error surface by examining two- or three-dimensional projection plots of the multidimensional error surface.[57,58] These plots can be used to verify whether a global minimum has been found and to probe parameter correlation.[45,61]

Example: Binding of Protein Kinase R to RNA

Protein kinase R (PKR) contains an N-terminal double-stranded RNA-binding domain (dsRBD) and a C-terminal kinase domain.[62] PKR binds to RNA in a nonspecific manner. Here, we characterize binding of the N-terminal dsRBD domain (amino acids 1–184) with a 20-base pair RNA by sedimentation equilibrium. Previous experiments established a binding stoichiometry of three dsRBD/RNA.[16] Based on these data, samples for sedimentation equilibrium were prepared at multiple concentrations of A (RNA) and B (protein) ([A] = 0.5 μM, [B] = 0.5, 1, and 2 μM). Data were collected at three detection wavelengths (230, 260, and 280 nm) and were globally analyzed using several alternative models to a model that incorporated AB, AB_2, and AB_3 species.[16] Figure 3 shows the data from all nine channels and a global fit to a model using three independent stepwise equilibrium constants. The best-fit parameters and statistics are summarized in Table II. The global fit is a good description of the experimental data, with no systematic deviations in the residuals and a low value of the RMS deviation of 0.00437 OD, which is consistent with the noise level in the optical system. The high value of ln K_1 of 18.32 indicates that first dsRBD binds strongly to the dsRNA (K_d = 11 nM). Although the 95% joint confidence intervals are relatively broad, there is a clear trend of decreasing binding strength with successive ligands, such that the third dsRBD binds significantly weaker, with ln K_3 = 14.07 (K_d = 0.78 μM). The ratios of the equilibrium constants are K_1/K_2 = 19 and K_1/K_3 = 70.

A decrease in equilibrium constants with successive ligand–binding events is predicted from statistical effects in the context of both the simple finite lattice model as well the finite lattice model that includes ligand overlap. Table II summarizes fits to several of these models in which the

[61] P. Schuck, C. G. Radu, and E. S. Ward, *Mol. Immunol.* **36**, 1117 (1999).
[62] M. J. Clemens and A. Elia, *J. Interferon Cytokine Res.* **17**, 503 (1997).

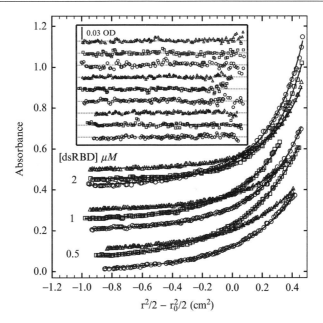

FIG. 3. Multiwavelength sedimentation equilibrium of PKR dsRBD binding to 20-mer dsRNA. The data were obtained under the following conditions: rotor speed, 23,000 rpm; temperature, 20°; RNA concentration, 0.5 μM; and protein concentrations of 0.5, 1, and 2 μM in 75 mM NaCl, 20 mM HEPES, 5 mM MgCl$_2$, 0.1 mM EDTA, pH 7.5. Detection wavelengths are 230 nm (O), 260 nm (□), and 280 nm (△). Solid lines are a global fit of the data to an unconstrained model of three ligands binding to the 20-mer RNA. The results of the fit are given in Table I. Inset: residuals. Traces have been vertically offset for clarity.

intrinsic equilibrium constant ln k is the fitted parameter and the values of ln K_1, ln K_2, and ln K_3 are calculated using Eqs. (6) and (11). In all cases, the stoichiometry s was fixed at 3. In the context of the model that includes ligand overlap, an equally good fit is found for $N = 14$ with $\Delta = 3$ (Table II); successively worse fits are found as N is reduced and Δ is held constant (data not shown). We have not considered larger site sizes, since for $\Delta = 3$ a value of $N > 14$ reduces S to 2. A good fit is also found for $N = 12$ with $\Delta = 4$. Again, worse fits are found with smaller site sizes and larger site sizes reduce the stoichiometry to 2. Models with $\Delta = 2$ are not considered because this close overlap between adjacent bound ligands would likely be disallowed due to steric hindrance. In summary, the multiwavelength sedimentation data fit well to several models where three dsRBD interact with the 20-bp dsRNA and the relative values of the three equilibrium constants are constrained according to the finite lattice model including ligand overlap. The best fits consistent with available structural information include a

TABLE II
EQUILIBRIUM CONSTANTS FOR BINDING OF PKR dsRBD TO A 20-MER RNA AS DETERMINED BY SEDIMENTATION EQUILIBRIUM

Model	$\ln K_1{}^a$	$\ln K_2{}^a$	$\ln K_3{}^a$	$\ln k^b$	RMS $\times 10^{-3\,c}$
Unconstrained[d]	18.32 [17.21, 19.48]	15.38 [14.43, 16.94]	14.07 [13.54, 14.47]	—	4.37
$N=14$, $\Delta=3$[e]	17.96	16.37	13.71	16.01 [15.77, 16.23]	4.48
$N=12$, $\Delta=4$[e]	18.51	16.82	13.60	16.31 [16.07, 16.58]	4.48

[a] Natural logarithm of the macroscopic binding constant. The values in brackets represent the 95% joint confidence intervals.
[b] Natural logarithm of the intrinsic binding constant. The values in brackets represent the 95% joint confidence intervals.
[c] Root mean square deviation of the fit in absorbance units (OD).
[d] Independent binding of three ligands. The natural logarithms of the macroscopic binding constants are the fitted parameters.
[e] Finite lattice models. The natural logarithms of the intrinsic binding constants are the fitted parameters. The macroscopic binding constants are calculated using coefficients determined by Eq. (11).

minimum offset (Δ) of 3–4 and site size (N) of 12–14 bp. These fits also reveal a value of the intrinsic binding constant of ln k = 16.0–16.3, or K_d = 83–110 nM.

It is worthwhile to graphically analyze the fitting results by plotting the modeled radial distributions for each species in the three samples in order to appreciate why the sedimentation equilibrium experiment is capable of resolving the three equilibrium constants. At the lowest protein concentration (Fig. 4A) the most abundant species are the AB complex and free A, along with low concentrations of AB$_2$ and free B. Thus, each of the species

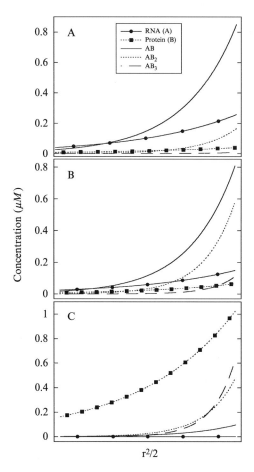

FIG. 4. Species distribution from sedimentation equilibrium analysis of PKR dsRBD binding to 20-mer dsRNA. Best fit parameters from the unconstrained analysis in Table II were used to generate radial concentration gradients for all the species present in solution. (A) [RNA] = 0.5 μM, [Protein] = 0.5 μM. (B) [RNA] = 0.5 μM, [Protein] = 1 μM. (C) [RNA] = 0.5 μM, [Protein] = 2 μM.

participating in the equilibrium governed by K_1 are populated: A, B, and AB. At intermediate protein concentration (Fig. 4B), the AB and AB_2 species are most prominent and at the highest protein concentration (Fig. 4C), the most abundant complexes are AB_2 and AB_3 along with high concentrations of free B. Thus at higher protein concentrations, the species governed by K_2 and K_3 become populated. To accurately define equilibrium constants it is necessary that all of the species participating in the equilibrium are populated. In the present case, it was necessary to span a range of A:B ratios to achieve significant population of each of the complexes. In the case of heterogeneous associations it has also been pointed out that much useful information can come from samples that are prepared far from the preferred stoichiometry.[25]

Conclusion

A variety of biophysical techniques with a range of capabilities are available to probe heterogeneous macromolecular interactions. The choice of technique is governed chiefly by the type of information that is required and by the specific properties of the material under investigation. Analytical ultracentrifugation measurements are particularly useful at the beginning of a study to define an association model; subsequently, higher-throughput methods may be more convenient to analyze large numbers of samples.

Acknowledgments

I thank Jason Ucci for his contributions to the studies of PKR and Jeff Lary for careful reading of this manuscript. This work was supported by the University of Connecticut's Research Advisory Council Programs.

[14] Estimation of Weights for Various Methods
of the Fitting of Equilibrium Data from the
Analytical Ultracentrifuge

By Marc S. Lewis and Michael M. Reily

Introduction

For approximately the past three decades, the direct fitting of concentration distribution as a function of radial position to an appropriate mathematical model has been the method of choice for equilibrium ultracentrifugal analysis. A variety of mathematical packages, both commercial

METHODS IN ENZYMOLOGY, VOL. 384   0076-6879/04 $35.00

such as MLAB, MATLAB, and Igor, and free, through a variety of sources, principally the RASMB and Beckman-Coulter, have been available to investigators: each has its advantages and disadvantages, and each has its partisans. The authors are partial to MLAB, principally because of familiarity coming from decades of use and because MLAB is so flexible, we have been able to do whatever we needed to do with it; we are sure that a number of other software packages would have done equally well; we simply have not had the time to try them. All of these packages utilize some form of either the Levenberg–Marquardt or the Simplex algorithm to obtain optimal values of the parameters by minimizing the appropriately weighted sum of squares of the residuals. Thus, if y is the dependent variable and is a function of the value of the independent variable x and the parameter values p_1, p_2, etc., and the standard deviation of the ith point is σ_i, then the parameter values are varied so that

$$\sum_{i=1}^{n} [(y_i - f(x_{i,p_1,p_2,...}))]^2 / \sigma_i^2 = \min \qquad (1)$$

It can be seen that the weight of each ith squared difference is the reciprocal of the square of σ_i, which is also the reciprocal of the variance of the ith point.

We have tried two methods for obtaining these weights. The first of these was to use the data from the third column of absorbance data, which is the value of σ for that absorbance as a function of radius, square it to obtain the variance, and use its reciprocal as the appropriate weight. The second of these was to use the MLAB EWT function, which performs a weighted moving spline fit on the absorbance data and takes the reciprocal of the square of the deviation of the fit point from the actual absorbance at each radial position as the weight. In each case the weights were normalized to unity using

$$W_{i,\text{normalized}} = nW_i / \sum_{i=1}^{n} W_i \qquad (2)$$

where n is the total number of the individual weights, W_i. This was to permit comparing unweighted and weighted fits. On the basis of the criteria of minimum sum of squares, minimum estimated parameter errors, and optimal distribution of residuals, we feel that the second method for weighting yielded better results and it has been our method of choice.

The function $f(x)$ will depend upon the mathematical model that is to be fit to the data. The simplest form is for an ideal monomeric component where the model then has the form

$$c(r) = c_b \exp[AM(r^2 - r_b^2)] + \varepsilon \qquad (3)$$

Since this work deals only with data from the absorbancy optical system, the cs are concentrations in those units; they may be converted to molarities by means of molar extinction coefficients; r_b is a reference radius, here taken to be the radius of the cell bottom, but can be any cell radial position with a corresponding value of c; M is the molar mass; ε is the baseline offset; and $A = (\partial\rho/\partial c)_\mu \omega^2/2RT$. $(\partial\rho/\partial c)_\mu$ is the density increment at constant chemical potential and is normally obtained experimentally by measuring the density of the solution in g cm^{-3} of the macromolecule dialyzed against the buffer as a function of the concentration of the macromolecule in g cm^{-3} based on its dry weight. Even if a relatively large quantity of the solute macromolecule is available, this can be a particularly difficult quantity to obtain experimentally because of the difficulty in obtaining an accurate dry weight of the solute and it is usually replaced by the approximation $(1 - \bar{v}\rho)$, where $\bar{v}$ is the partial specific volume of the macromolecule and ρ is the density of the solvent. The partial specific volume is usually calculated from the partial specific volumes of the individual amino acid residues composing the protein, and thus is considered to be a compositional value. This is a reasonable approximation if the solvent is dilute, but it is an approximation based upon the assumption of a two-component system and it is not a thermodynamically defined value. It is, however, the most commonly used approach. ω is the angular velocity of the rotor in radians/s, R is the gas constant, and T is the absolute temperature. Since better criteria than ultracentrifugal homogeneity are now in common use and amino acid sequence and mass spectroscopy are now more precise measures of molar mass, a better use of this model with ideal homogeneous solutes is for measuring the reduced mass, $M(\partial\rho/\partial c)_\mu$, which appears to be superior to the direct densitometric measurement of $(\partial\rho/\partial c)_\mu$ for obtaining thermodynamically valid values of this quantity. The use of this method has been the practice in this laboratory for the past decade for such cases where it is applicable.

The mathematical model given in Eq. (2) may be readily extended to other models. For the case of a mixture of components it will have the form

$$c(r) = \sum_{i=1}^{n} c_{b_i}\exp\left[A_iM_i(r^2 - r_b^2)\right] + \varepsilon \tag{4}$$

It is readily apparent that the values of the partial specific volumes in the various values of A_i will be approximations at best and these uncertainties will be reflected in the values of the molar masses.

Of greatest interest now is the use of sedimentation equilibrium for the study of molecular associations. The two most commonly seen associating systems are the homogeneous associations such as monomer–dimer, monomer–trimer, and monomer–dimer–trimer, and the heterogeneous associations such as the $A_i + B_j \Leftrightarrow A_iB_j$, with a variety of stoichiometries depending on the values of i and j, where these may have the same value or may differ. Homogeneous associations may be described by the model

$$c(r) = c_{b,1}\exp\left[AM(r^2 - r_b^2)\right] + \sum_{i=2}^{n} c_b^i\exp\left[\ln K_{1,i} - \ln E_{1,i} + iAM(r^2 - r_b^2)\right] + \varepsilon$$

(5)

In this model, all of the molar equilibrium constants, $K_{1,i}$, are based on the concept of the oligomer being formed from the monomer. This has proved to be the most effective way of computationally estimating equilibrium constants, and in multicomponent systems, the intermediate equilibrium constants are readily calculated from the monomer–i-mer values. The term $E_{1,i}$ is used to convert the molar equilibrium constant to the actual measuring concentration scale and for the absorbance optical system has the value $E_{1,i} = E_1^{i-1}/i$, where E_1 is the molar extinction coefficient of the monomer at the scanning wavelength. A simple 1:1 stoichiometry heterogeneous association can be described by the model

$$\begin{aligned} c(r) = &\, c_{b,A} \exp(A_A M_A \delta r^2) + c_{b,B} \exp(A_B M_B dr^2) \\ &+ c_{b,A}c_{b,B} \exp\left[\ln K_{AB} - \ln E_{AB} + (A_A M_A + A_B M_B)\delta r^2\right] + \varepsilon \end{aligned}$$

(6)

where $\delta r^2 = (r^2 - r_b^2)$. This model assumes that there is no volume change upon association and that the reduced mass of the complex is equal to the sum of the reduced masses of the reactants. The extinction coefficient term, E_{AB}, is defined as $E_{AB} = E_A E_B/(E_A + E_B)$. This definition must be appropriately modified when the stoichiometry is other than 1:1. The 1:1 stoichiometry type of associating system is a perfect example of where the reduced masses of the reactants to be used in the modeling can be optimally determined by measuring their reduced mass in separate experiments. The associating systems described previously are a minimal example of what is potentially possible, although with increasing difficulty and decreasing accuracy for the determination of the values of ln Ks as the system becomes more complex. An attempt to improve this accuracy for all associating systems led to the development of the use of intensities rather than absorbancies. At the time this work was done, the alternative to analyzing intensities appeared to be the use of a maximum likelihood estimator (MLE) method. This appeared to be computationally more difficult and complex than analyzing intensities, so it was not attempted.

Equilibrium Analyses by Means of Intensities

The absorbtion optical system of either the XL-A or the XL-I analytical ultracentrifuges operates essentially like any other dual-beam spectrophotometer in that it measures the intensities of light passing through the channels of a double-sector centerpiece and compares the intensities of light passing through the channel containing the solvent with the intensities of light passing through the channel containing the solute(s) of interest. Since the light passing through the reference channel is not a true incident intensity, but is attenuated by the windows and the solvent, we have chosen to denote its intensity as I_0. Similarly, the light passing through the solution of interest differs only by virtue of the light absorbed by the solute molecules, so we denote its intensity as I_S. Then, the absorbance is denoted as $\log_{10}(I_0/I_S)$. Since, as demonstrated by Dimitriadis and Lewis,[1,2] both of these intensities have normally distributed (Gaussian) error, the error of their ratio must be a Cauchy distribution and the error distribution of the absorbancies must be a logarithmically skewed Cauchy distribution. Thus, since fitting data by means of nonlinear least-squares explicitly assumes that the error distribution must be Gaussian or, suboptimally, unknowable, fitting analytical ultracentrifuge absorption data by nonlinear least squares is theoretically incorrect. The senior author (M. S. L.) is at least as guilty of this as any investigator in the field. It is fortunate that the error so induced appears to be relatively small in almost all cases, so that we do not have to question the essential validity of a very large volume of significant research. However, as we deal with increasingly complex systems, the possibility of significant error increases and it becomes increasingly important to have improved analytical methodology. Thus, while there is also an aesthetic satisfaction in using a more appropriate mathematical method, it must, of course, be tempered by the pragmatic consideration of whether or not the gain is worth the additional cost in time and effort.

Two approaches to the use of intensities were developed by Dimitriadis and Lewis. In the first of these, the values of $I_0(r)$ were smoothed by the use of moving cubic splines to give values we denote as $I_0'(r)$. This was necessary since the photomultiplier response as a function of time and radial position could not be modeled. Then, the values of $I_S(r)$ were fit by nonlinear least-squares using the model

$$I_S(r) = I_0'(r)10^{-c(r,p_{1,2,3,\ldots})} \tag{7}$$

where $c(r, p_{1,2,3}, \ldots)$ is the model of concentration distribution as a function of radial position and the fitting parameters for fitting absorbance data for that particular experiment. In the other approach, both intensities at each radial position were jointly fit by adding a column of zeros to the data matrix so that we now had an n-row matrix with four columns, radius, I_S,

[1] E. K. Dimitriadis and M. S. Lewis, *Prog. Colloid Polym. Sci.* **107,** 20 (1997).
[2] E. K. Dimitriadis and M. S. Lewis, *Methods Enzymol.* **321,** 121 (2000).

I_0, and zeros and then fitting for the zeros as a function of radial position and the two intensities using the model

$$G(r, I_{S,r}, I'_{0,r}) = I_{S,r} - I'_0 10^{-[c(r,p_{1,2,3,\ldots})]} \tag{8}$$

With perfect data using this model, the values of G at all radial positions should be zeros. With real data, the values of the fitting parameters, p_1, p_2, p_3,... are varied in the nonlinear least-squares fitting procedure to minimize the sum of the squares of the deviations of G from zero. These procedures were used with good results by Dimitriadis et al.[3]

It should be noted that all of these fits were unweighted; optimally, non-linear least-squares fits should have each data point weighted with the reciprocal of its variance. The problem now was to find a means of obtaining these variances. The first of the procedures did not appear to offer a ready solution to this problem. However, with a properly designed experiment, using the second procedure would permit measuring the individual variances of I_S and I_0, which were independent of each other, and then calculating the total variance at each radial position using

$$\text{var}(\text{total}, r) = \text{var}(I_S, r) + \text{var}(I_0, r)10^{-2[c(r,p_{1,2,3,\ldots})]} \tag{9}$$

The total weight at each radial position was then the reciprocal of the total variance at that radius. The weights were then normalized to unity as previously described so that unweighted and weighted fits could be compared.

To test this procedure we performed an equilibrium experiment in an XL-A ultracentrifuge with one cell containing a 5-mm column of bovine serum albumin (BSA) in phosphate-buffered saline (PBS) and run at 10,000 rev/min at 20°. After allowing 261 h for attainment of equilibrium we used the Methods option of the XL-A software to take four sets of 99 individual nonreplicated scans with radial increments of 0.050 cm over the next 6 h. The very large resulting data set was then used to obtain the mean values of the intensities and their variances for calculation of the normalized weights as a function of radial position. These data were then analyzed, both unweighted and weighted, using Eq. (8) with the molar mass, the reference concentration at the nominal cell bottom, and the baseline error as fitting parameters. These are presented in Table I. In this same experiment we then performed two 10-replicate scans with radial increments of 0.001 cm, one in the intensity mode and the other in the absorbance mode. We then analyzed the intensity scan by performing moving spline

[3] E. K. Dimitriadis, R. Prasad, M. K. Vaske, L. Chen, A. E. Tomkinson, M. S. Lewis, and S. H. Wilson, J. Biol. Chem. 273(32), 20540 (1998).

TABLE I
RESULTS OF NONLINEAR LEAST-SQUARES REGRESSION

Unweighted nonlinear least-squares fit of multiple intensity data sets:
  $M = 67529 \pm 4563$
  $c_b = 1.3724 \pm 0.0888$
  $\varepsilon = -0.00572 \pm 0.01120$
  Sum of squares = 5.337×10^5
Weighted nonlinear least-squares fit of multiple intensity data sets:
  $M = 61055 \pm 7772$
  $c_b = 1.2754 \pm 0.1914$
  $\varepsilon = -0.02419 \pm 0.01670$
  Sum of squares = 6.612×10^5
Unweighted nonlinear least-squares fit of single intensity data set:
  $M = 64927 \pm 892$
  $c_b = 1.2958 \pm 0.0148$
  $\varepsilon = -0.01119 \pm 0.00251$
  Sum of squares = 1.0632×10^7
Weighted nonlinear least-squares fit of single intensity data set:
  $M = 65524 \pm 422$
  $c_b = 1.3142 \pm 0.0046$
  $\varepsilon = -0.01305 \pm 0.00186$
  Sum of squares = 1.9448×10^6
Unweighted nonlinear least-squares fit of single absorbancy data set:
  $M = 65032 \pm 403$
  $c_b = 1.3029 \pm 0.0034$
  $\varepsilon = -0.01185 \pm 0.00187$
  Sum of squares = 4.7066×10^{-2}
Weighted nonlinear least-squares fit of single absorbancy data set:
  $M = 64848 \pm 283$
  $c_b = 1.3044 \pm 0.0023$
  $\varepsilon = -0.01489 \pm 0.00162$
  Sum of squares = 2.2059×10^{-2}

smoothing of each of the two intensities and used the square of the deviation of an actual intensity at a particular radius from its smoothed value as an approximation of its variance at that radius. These data were treated as described previously and the results are also presented in Table I. The time and labor involved in performing the mathematically rigorous analyses are excessive for any practical application, but serve as a benchmark for comparative purposes. The time required for the method of approximating the variances, while longer than that required for weighted nonlinear least-squares analysis, is reasonable. However, when we analyzed the absorbance data using Eq. (2) as a mathematical model for weighted nonlinear least-squares analysis, we obtained a better result, also shown in Table I.

Several important conclusions can be drawn from the data presented in Table I. The first is that the statistically rigorous multiple data set experiment yielded significantly inferior results. One possible cause for this might be the relatively large radial increments used when collecting data. Considering how long it took to acquire the data we used, the time required to acquire larger data sets with smaller radial increments and to process the resulting data would have been prohibitive. In contrast to this, with the exception of the sum of squares, the unweighted single intensity set analysis produced a markedly better result on the basis of the returned estimated parameter errors. The larger sum of squares simply reflects the fact that the actual data set analyzed here had many more points than were in the multiscan set where averages at each radial position were used. Weighting the single intensity data set using estimated variances effected a significant improvement both in returned parameter error estimates and in the sum of squares. The results for the single unweighted absorbancy data set do not differ from the weighted intensity results in any significant way. The sums of squares of the two sets cannot be compared because one was for intensities and the other for absorbancies. The results from the weighted absorbancy data clearly indicate an advantage for weighting using estimated variances. The returned estimated parameter errors and the sum of squares all show significant reduction.

This was not a totally unexpected result as we had previously observed a similar result with simulated data using the second weighting procedure. While we have obtained better results with unweighted intensity analysis for other systems, particularly those involving associations, these failures with both simulated and real data in addition to the time and effort required for this type of analysis create a very definite uncertainty with regard to the general applicability of intensity methods to equilibrium analysis. In view of this we have now considered the use of robust regression of absorbance data as a possible answer.

Robust Regression

The simplest form of robust regression is designated as L-1 regression and is quite similar to least-squares regression, differing in that it minimizes the sum of the absolute deviations rather than the sum of the squares of the deviations and uses the value of σ rather than σ^2 of each point as its weight. This is described by

$$\sum_{i=1}^{n} \text{abs}[(y_i - f(x_{i,p_1,p_2,...}))]/\sigma_i = \min \tag{10}$$

L-1 regression offers two significant advantages for the analysis of absorbance data. The first is that outliers in the data have significantly less effect on the fit obtained.[4] Gross outliers in scans that are the result of failure of firing of the flash lamp give easily detected spikes that are easily and appropriately deleted. However, there may be significantly more subtle outliers as a result of window aberrations, smudges, lint, etc., and these can present more difficulty in objective editing. L-1 regression significantly reduces the effects of these on the quality of the fit. Of greater significance in fitting absorbance data, which has a logarithmically skewed Cauchy error distribution, is that L-1 regression is considered a preferred method for the fitting of data, which has so-called "fat-tailed" error distributions, of which the Cauchy distribution is given as a prime example.[5] By inference, the logarithmically skewed Cauchy distribution is also "fat-tailed." A further advantage for ultracentrifugal analysis is that, as noted before, multireplicate scans have values of σ as a function of radial position presented in the third data column. Thus, it is very easy to obtain proper weights by taking the reciprocals of these values and normalizing to unity as described previously. We analyzed the absorbance data referred to earlier by this method and obtained results that we consider significantly better than those from nonlinear least-squares analysis. The results are compared in Table II.

Examination of the data presented in Table II leads to some fairly obvious conclusions. Unweighted and weighted L-1 regression appear to give significantly reduced returned estimated parameter errors when compared to unweighted and weighted nonlinear least-squares regression. The sum of squares cannot be compared to the sum of residuals for obvious reasons. While weighting does not appear to effect a significant improvement for the L-1 regression shown here, we feel that it probably should be done routinely for any added benefits it might confer and also as a matter of mathematical rigor. Since it is so easy to do, there is little reason not to do so. An illustration of the fit and the distribution of the residuals is presented in Fig. 1.

Discussion

Our experiences, as related here, lead to some reasonable conclusions. The first of these is that while the nonlinear least-squares analysis of intensity data is more rigorous mathematically than similar analysis of absorbancy data, its potential unreliability, which we cannot explain, and the

---

[4] M. L. Johnson, *Methods Enzymol.* **321,** 417 (2000).

[5] R. D. Armstrong and P. O. Beck, "Robust Regression" (K. D. Lawrence and J. L. Arthur, eds.), p. 89. Marcel Decker, Inc., New York, 1990.

TABLE II
COMPARISON OF PN-L LS REGRESSION WITH L-1 REGRESSION

Unweighted nonlinear least-squares fit of absorbancy data:
   $M = 65032 \pm 403$
   $c_b = 1.3029 \pm 0.0034$
   $\varepsilon = -0.01185 \pm 0.00187$
   Sum of squares $= 4.7066 \times 10^{-2}$
Weighted nonlinear least-squares fit of absorbancy data:
   $M = 64848 \pm 283$
   $c_b = 1.3044 \pm 0.0023$
   $\varepsilon = -0.01489 \pm 0.00162$
   Sum of squares $= 2.2059 \times 10^{-2}$
Unweighted L-1 fit of absorbancy data:
   $M = 64881 \pm 120$
   $c_b = 1.3045 \pm 0.00061$
   $\varepsilon = -0.01467 \pm 0.00072$
   Sum of residuals $= 2.943$
Weighted L-1 fit of absorbancy data:
   $M = 64918 \pm 117$
   $c_b = 1.3046 \pm 0.00091$
   $\varepsilon = -0.01446 \pm 0.00078$
   Sum of residuals $= 2.841$

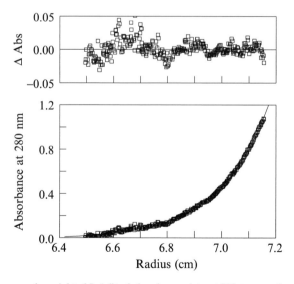

FIG. 1. Lower graph: weighted L-1 fit of absorbance data at 280 nm as a function of radial position for BSA in PBS at 20° and 10,000 rev/min. The fitting parameters are presented in Table II. Upper graph: distribution of the residuals for this fit.

much greater effort required to use it properly make it less useful than we consider justified. In contrast to this, L-1 robust regression analysis of absorbancy data has a sounder mathematical basis than nonlinear least-squares regression of the same data and has other attractive properties, discussed earlier. Also, based so far on relatively limited experience, it appears to give superior results. Two questions remain. The first is the estimation of parameter errors. We have presented estimated values returned when performing L-1 regression using MLAB. We are working on ways of validating these or of obtaining more reliable estimates. At this time we regard what is known as a balanced bootstrap simulation as a promising method.[6] In this procedure the resampling is performed in such a way that each data point is present the same number of times in the union of all of the resamples. This ensures that the sample mean and the grand mean of the bootstrap resamples are identical and effect reduction in variance and mean squared error. The second question is the matter of software. For those who use MLAB, as we do, all that is required for L-1 regression is to type in the command "fitnorm=1" before beginning the fit. We do not know if comparable methods are possible with other software. However, work is under way on the development of suitable software for distribution from the RASMB site. We have performed nonlinear least-squares and L-1 regression analyses on several interacting systems and have had uniformly superior results to date with L-1 regression. In view of the fact that we must consider intensity analysis to be a significant disappointment that took some time and experience to discover, we are cautiously optimistic about the future of L-1 regression applied to equilibrium analysis. Since the performance of properly weighted L-1 regression is as rapid and is no more laborious or difficult than approximately weighted nonlinear least-squares regression, we would like to encourage other investigators to enlarge our body of experience with regard to its apparent superiority.

[6] P. Hall, "The Bootstrap and Edgeworth Expansion," p. 293. Springer-Verlag, New York, 1992.

[15] Applications of NMR Spin Relaxation Methods for Measuring Biological Motions

By GURUVASUTHEVAN R. THUDUPPATHY and R. BLAKE HILL

Introduction

The activities of biological macromolecules are typically dependent on their conformational flexibility. A recent example is the activation by blue light of the oat plant phototropin, a multidomain protein involved in regulating phototropism, stomatal opening, chloroplast relocation, and leaf opening.[1] Activation of the phototropin involves a kinase domain and arises from light-induced conformational changes in α-helix J of the preceding domain that are propagated across the domains to switch the kinase from an inactive (dark) to an active (light) state. This example illustrates the importance of conformational flexibility to biological activity, which has clearly been appreciated for some time,[2,3] however, little quantitative information on these dynamic processes is available at atomic resolution.

Nuclear magnetic resonance (NMR) spectroscopy is particularly well-suited to address this problem for two reasons. First, NMR experiments can provide atomic resolution of dynamic processes. The power to characterize dynamic processes of proteins at the atomic level by NMR spectroscopy was demonstrated in 1971 by Allerhand and co-workers.[4] They estimated the differences in the rotational correlation times of bovine pancreatic ribonuclease A in the folded and unfolded states by measuring NMR spin relaxation of the naturally abundant ^{13}C signal. These studies illustrated the ability of NMR to characterize biomolecular dynamics, but were limited in their general utility because of the difficulty in obtaining biological samples at the high concentrations required for detection of the ^{13}C signal at natural abundance. Now, however, NMR spin relaxation is widely applicable to biomolecular dynamics because of the ability to incorporate stable, NMR-active isotopes such as ^{15}N and ^{13}C, either uniformly or selectively, into biomolecules.[5–9] Thus, nearly every atom in a

[1] S. M. Harper, L. C. Neil, and K. H. Gardner, *Science* **301,** 1541 (2003).
[2] K. U. Linderstrom-Land and J. A. Schellman, *Enzyme* **1,** 443 (1959).
[3] K. Wuthrich and G. Wagner, *Experientia* **31,** 726 (1975).
[4] A. Allerhand, D. Doddrell, V. Glushko, D. W. Cochran, E. Wenkert, P. J. Lawson, and F. R. N. Gurd, *J. Am. Chem. Soc.* **93,** 544 (1971).
[5] J. Marley, M. Lu, and C. Bracken, *J. Biomol. NMR* **20,** 71 (2001).

biological macromolecule could potentially serve as a probe for dynamic processes. Additionally, the NMR-active nuclei (^{1}H, ^{15}N, ^{13}C, ^{31}P) do not introduce perturbations to the structure or dynamics.[10]

The second reason that NMR is uniquely poised to provide quantitative information on dynamic processes is the wide range of time scales (from picoseconds to several hours) that is accessible by NMR methods. The application of NMR methods for quantitative descriptions of biomolecular dynamic processes that occur on the picosecond to nanosecond time scale is well known.[10–18] The physical basis for these experiments is that molecular motions cause weak, fluctuating magnetic fields that result in magnetization decay (or recovery) commonly referred to as spin relaxation. In principle, by measuring the rate constants for spin relaxation (R_1, R_2, and cross-relaxation), the magnitude and frequency of molecular motions can be extracted.[10–12,15,19] While such spin relaxation data analysis is by no means trivial, the application of these methods to both proteins and nucleic acids has become routine.[13,17,18,20,21] These studies have identified that the backbone of proteins is fairly rigid as expected. Furthermore, in some cases, such experiments have led to the identification of flexible regions of biomolecules important for function. In HIV-1 protease, for example, two symmetric flaps of the enzyme have been shown to undergo motions on the picosecond to nanosecond time scale that are thought to facilitate substrate binding and product release.[22,23]

[6] A. L. Lee, J. L. Urbauer, and A. J. Wand, *J. Biomol. NMR* **9,** 437 (1997).

[7] M. Rance, P. E. Wright, B. A. Messerle, and L. D. Field, *J. Am. Chem. Soc.* **109,** 1591 (1987).

[8] N. K. Goto and L. E. Kay, *Curr. Opin. Struct. Biol.* **10,** 585 (2000).

[9] D. M. Lemaster, *Prog. NMR Spectrosc.* **26,** 371 (1994).

[10] L. E. Kay, *Biochem. Cell Biol. Biochim. Biol. Cell* **76,** 145 (1998).

[11] A. G. Palmer, *Annu. Rev. Biophys. Biomol. Struct.* **30,** 129 (2001).

[12] K. T. Dayie, G. Wagner, and J. F. Lefevre, *Annu. Rev. Phys. Chem.* **47,** 243 (1996).

[13] J. W. Peng and G. Wagner, *Methods Enzymol.* **239,** 563 (1994).

[14] D. M. LeMaster and D. M. Kushlan, *J. Am. Chem. Soc.* **118,** 9255 (1996).

[15] D. Idiyatullin, V. A. Daragan, and K. H. Mayo, *J. Magn. Reson.* **161,** 118 (2003).

[16] N. Tjandra, A. Szabo, and A. Bax, *J. Am. Chem. Soc.* **118,** 6986 (1996).

[17] K. L. Mayer, M. R. Earley, S. Gupta, K. Pichumani, L. Regan, and M. J. Stone, *Nat. Struct. Biol.* **5,** 5 (2003).

[18] K. L. Mayer and M. J. Stone, *Proteins* **50,** 184 (2003).

[19] R. E. London, *Methods Enzymol.* **176,** 358 (1989).

[20] A. G. Palmer, *Curr. Opin. Struct. Biol.* **7,** 732 (1997).

[21] R. Ishima and D. A. Torchia, *Nat. Struct. Biol.* **7,** 740 (2000).

[22] L. K. Nicholson, T. Yamazaki, D. A. Torchia, S. Grzesiek, A. Bax, S. J. Stahl, J. D. Kaufman, P. T. Wingfield, P. Y. S. Lam, P. K. Jadhav, C. N. Hodge, P. J. Domaille, and C. H. Chang, *Nat. Struct. Biol.* **2,** 274 (1995).

Additionally, NMR spin relaxation methods have revealed that pico-second to nanosecond time scale motions contribute to the thermodynam-ics of biomolecular folding and structure.[24–33] For example, Akke *et al.*[34] developed an analytical expression that relates the changes in backbone mobility derived from NMR spin relaxation data to the Gibbs free energy of a process. The approach was applied to the cooperative Ca^{2+} binding to the protein calbindin D_{9k} in which significant contributions to the free energy of cooperativity were demonstrated to arise from motions of the protein backbone on the nanosecond to picosecond time scale. Another important example was provided by Yang and Kay to the folding–unfolding transition of an SH3 protein domain. In this work, they demon-strated picosecond to nanosecond time scale changes in backbone motion correlate with changes in the conformational entropy of folding.[35]

The NMR methods and applications directed at measuring such pico-second to nanosecond time scale motions have been the subject of a number of thorough reviews.[10,21,36–38] While the majority of these studies have focused on protein backbone motions, NMR spin relaxation methods also provide the ability to measure protein side chain dynamics.[16,39–44] Lee and Wand have used ^2H NMR spin relaxation experiments to probe the

[23] T. Yamazaki, A. P. Hinck, Y. X. Wang, L. K. Nicholson, D. A. Torchia, P. Wingfield, S. J. Stahl, J. D. Kaufman, C. H. Chang, P. J. Domaille, and P. Y. Lam, *Protein Sci.* **5,** 495 (1996).

[24] L. Spyracopoulos and B. D. Sykes, *Curr. Opin. Struct. Biol.* **11,** 555 (2001).

[25] F. A. Mulder, A. Mittermaier, B. Hon, F. W. Dahlquist, and L. E. Kay, *Nat. Struct. Biol.* **8,** 932 (2001).

[26] D. W. Yang, D. Shortle, D. R. Muhandiram, J. D. Forman-Kay, and L. E. Kay, *Biophys. J.* **74,** A137 (1998).

[27] L. E. Kay, D. R. Muhandiram, G. Wolf, S. E. Shoelson, and J. D. Forman-Kay, *Nat. Struct. Biol.* **5,** 156 (1998).

[28] K. H. Gardner and L. E. Kay, *Annu. Rev. Biophys. Biomol. Struct.* **27,** 357 (1998).

[29] C. Bracken, P. A. Carr, J. Cavanagh, and A. G. Palmer, *J. Mol. Biol.* **285,** 2133 (1999).

[30] M. J. Stone, *Acc. Chem. Res.* **34,** 379 (2001).

[31] M. J. Stone, S. Gupta, N. Snyder, and L. Regan, *J. Am. Chem. Soc.* **123,** 185 (2001).

[32] M. J. Seewald, K. Pichumani, C. Stowell, B. V. Tibbals, L. Regan, and M. J. Stone, *Protein Sci.* **9,** 1177 (2000).

[33] L. Zidek, M. V. Novotny, and M. J. Stone, *Nat. Struct. Biol.* **6,** 1118 (1999).

[34] M. Akke, R. Bruschweiler, and A. G. Palmer, *J. Am. Chem. Soc.* **115,** 9832 (1993).

[35] D. W. Yang and L. E. Kay, *J. Mol. Biol.* **263,** 369 (1996).

[36] V. A. Daragan and K. H. Mayo, *Prog. NMR Spectrosc.* **31,** 63 (1997).

[37] M. W. F. Fischer, A. Majumdar, and E. R. P. Zuiderweg, *Prog. NMR Spectrosc.* **33,** 207 (1998).

[38] J. G. Kempf and J. P. Loria, *Cell Biochem. Biophys.* **37,** 187 (2003).

[39] L. K. Nicholson, L. E. Kay, D. M. Baldisseri, J. Arango, P. E. Young, A. Bax, and D. A. Torchia, *Biochemistry* **31,** 5253 (1992).

temperature-dependent dynamics of methyl-containing side chains of the Ca^{2+}-signalling protein calmodulin in association with a target peptide.[45,46] They found three classes of motion that were suggestive of the heterogeneous glass-like nature of proteins proposed by Frauenfelder and co-workers.[47–49] Furthermore, the temperature dependence of the methyl dynamics of the calmodulin–peptide complex revealed a heterogeneous distribution of motions that suggests local contributions to the heat capacity of binding. Subsequent molecular dynamics simulations have corroborated these results.[50] Thus, NMR coupled with molecular dynamics simulations is a powerful tool for exploring the energy landscape of fast biomolecular processes.[11,50–53]

Many important biological processes such as allosteric transitions, enzyme catalysis, and protein folding occur on the microsecond to millisecond time scale.[54,55] These so-called slower time scale motions have been known for at least 45 years to affect the observed chemical shift and lineshape of the NMR signal.[56] Theoretical and experimental approaches were subsequently developed to measure these processes in small molecules, including motions, that arise from chemical kinetics.[57–64] More recently, developments in heteronuclear (^{13}C and ^{15}N) NMR techniques have led to

[40] L. K. Nicholson, L. E. Kay, F. Delaglio, A. Bax, and D. A. Torchia, *Biophys. J.* **64**, A182 (1993).

[41] A. J. Wand, J. L. Urbauer, R. P. McEvoy, and R. J. Bieber, *Biochemistry* **35**, 6116 (1996).

[42] D. M. LeMaster, *J. Am. Chem. Soc.* **121**, 1726 (1999).

[43] P. F. Flynn, R. J. Bieber, J. L. Urbauer, H. Zhang, A. L. Lee, and A. J. Wand, *Biochemistry* **40**, 6559 (2001).

[44] K. H. Lee, D. Xie, E. Freire, and L. M. Amzel, *Proteins* **20**, 68 (1994).

[45] A. L. Lee, S. A. Kinnear, and A. J. Wand, *Nat. Struct. Biol.* **7**, 72 (2000).

[46] A. L. Lee and A. J. Wand, *Nature* **411**, 501 (2001).

[47] H. Frauenfelder, G. A. Petsko, and D. Tsernoglou, *Nature* **280**, 558 (1979).

[48] H. Frauenfelder, F. Parak, and R. D. Young, *Annu. Rev. Biophys. Biophys. Chem.* **17**, 451 (1988).

[49] H. Frauenfelder, S. G. Sligar, and P. G. Wolynes, *Science* **254**, 1598 (1991).

[50] N. V. Prabhu, A. L. Lee, A. J. Wand, and K. A. Sharp, *Biochemistry* **42**, 562 (2003).

[51] V. Tsui, I. Radhakrishnan, P. E. Wright, and D. A. Case, *J. Mol. Biol.* **302**, 1101 (2000).

[52] V. A. Daragan and K. H. Mayo, *Biochemistry* **32**, 11488 (1993).

[53] P. E. Smith, R. C. Vanschaik, T. Szyperski, K. Wuthrich, and W. F. Vangunsteren, *J. Mol. Biol.* **246**, 356 (1995).

[54] S. Kumar, B. Ma, C. J. Tsai, H. Wolfson, and R. Nussinov, *Cell Biochem. Biophys.* **31**, 141 (1999).

[55] V. Daggett and A. Fersht, *Nat. Rev. Mol. Cell. Biol.* **4**, 497 (2003).

[56] H. M. McConnell, *J. Chem. Phys.* **28**, 430 (1958).

[57] J. Jen, *J. Magn. Reson.* **30**, 111 (1978).

[58] J. T. Gerig and A. D. Stock, *Org. Magn. Reson.* **7**, 249 (1975).

[59] J. P. Carver and R. E. Richards, *J. Magn. Reson.* **6**, 89 (1972).

[60] Z. Luz and S. Meiboom, *J. Chem. Phys.* **39**, 366 (1963).

experiments that have greatly extended the utility of the approaches in two regards. First, the newer methods are typically applied to biomolecules uniformly labeled with ^{15}N or ^{13}C providing dynamics data for each labeled atom. Second, these newer methods significantly extend the range of measurable time scales.[65–68] For processes on the fast microsecond time scale ($\sim10^3$–10^6 s^{-1}), the transverse relaxation rate will be affected and can be measured using dynamic lineshape analysis[69] or $R_{1\rho}$ experiments.[70] For example, Raleigh and co-workers recently reported the folding rate constant for the 36-residue villin headpiece subdomain to be 0.5–2.0 $\times$ 10^5 s^{-1} from dynamic lineshape analysis.[71] In another important application that laid the foundation for subsequent work, Davis *et al.* measured the dissociation rate constant for tubercidin, an inhibitor of the enzyme purine nucleoside phosphorylase from *Escherichia coli,* to be 2400 s^{-1} using $R_{1\rho}$ experiments.[62]

Motions on the microsecond to millisecond time scale ($\sim10^6$–10^3 s^{-1}) increase the transverse relaxation rate, R_2. These motions can be detected indirectly during the spin relaxation data analysis using the model free approach introduced by Lipari and Szabo.[72,73] In this manner, Volkman *et al.* measured backbone motions of the phosphorylation-driven signaling protein NtrC in three functional states: the inactive unphosphorylated[74] state, the phosphorylated active state, and a mutant NtrC that is in a partially active state.[75] The comparison of the backbone motions for these states led to the conclusion that microsecond to millisecond time scale motions correlated with activation by phosphorylation. This work is particularly significant in the demonstration that a single-domain signaling protein,

[61] T. J. Swift and R. E. Connick, *J. Chem. Phys.* **37,** 307 (1962).

[62] D. G. Davis, M. E. Perlman, and R. E. London, *J. Magn. Reson. B* **104,** 266 (1994).

[63] A. Allerhand and H. S. Gutowsky, *J. Chem. Phys.* **42,** 1587 (1965).

[64] A. Allerhand and H. S. Gutowsky, *J. Chem. Phys.* **41,** 2115 (1964).

[65] J. P. Loria, M. Rance, and A. G. Palmer, *J. Am. Chem. Soc.* **121,** 2331 (1999).

[66] R. Ishima and D. A. Torchia, *J. Biomol. NMR* **25,** 243 (2003).

[67] C. Y. Wang, M. J. Grey, and A. G. Palmer, *J. Biomol. NMR* **21,** 361 (2001).

[68] R. Ishima and D. A. Torchia, *J. Biomol. NMR* **14,** 369 (1999).

[69] G. S. Huang and T. G. Oas, *Proc. Natl. Acad. Sci. USA* **92,** 6878 (1995).

[70] J. W. Peng and G. Wagner, *Biochemistry* **31,** 8571 (1992).

[71] M. Wang, Y. Tang, S. Sato, L. Vugmeyster, C. J. McKnight, and D. P. Raleigh, *J. Am. Chem. Soc.* **125,** 6032 (2003).

[72] G. Lipari and A. Szabo, *J. Am. Chem. Soc.* **104,** 4559 (1982).

[73] G. Lipari and A. Szabo, *J. Am. Chem. Soc.* **104,** 4546 (1982).

[74] G. M. Clore, A. Szabo, A. Bax, L. E. Kay, P. C. Driscoll, and A. M. Gronenborn, *J. Am. Chem. Soc.* **112,** 4989 (1990).

[75] B. F. Volkman, D. Lipson, D. E. Wemmer, and D. Kern, *Science* **291,** 2429 (2001).

NtrC, is capable of undergoing allosteric activation by a population-shift mechanism and not an induced-fit mechanism.

More recently, Loria and co-workers have shown a correlation between catalysis and backbone motion for ribonuclease A (RNase A) by directly measuring the contributions of microsecond to millisecond time scale motions to the transverse relaxation rates using Carr–Purcell–Meiboom–Gill (CPMG)-based experiments.[76] They found that the backbone motion was identical for all residues (k_{ex} 1640 s^{-1}) and was similar to the k_{cat} value for RNase A (1900 s^{-1}). Temperature-dependent CPMG-based experiments allowed extraction of activation energy for microsecond to millisecond motion and was found to be similar to the activation barrier for enzyme catalysis. For processes on the high millisecond to second time scale (0.1–10 s^{-1} for ^{15}N), ZZ-exchange spectroscopy can measure the affect of these motions on the transverse relaxation rate constant, R_2. Farrow et al. used ^{15}N ZZ-exchange experiments to measure the folding and unfolding rate constants for the drk SH3 domain.[77] These results clearly illustrate the utility of NMR-based methods in linking conformational flexibility to biological function.

This chapter presents data analysis considerations for these new methods that are directed at measuring conformational flexibility or chemical kinetics on the microsecond to millisecond time scale. While the data discussed in this review are primarily from CPMG-based experiments, the principles presented are also applicable to data collected by $R_{1\rho}$ relaxation and ZZ-exchange experiments, even though these latter cases are treated with a slightly different formalism. Several excellent reviews have recently appeared that detail theoretical and experimental aspects of all of these methods, in addition to recent applications.[38,78–81] While this chapter draws from examples in the literature applied to proteins, the principles presented are also applicable to other biomolecules.[82] The discussion is aimed at the biologist or biochemist without formal NMR training, however, our goal is that specialists will also benefit from the data analysis considerations.

[76] R. Cole and J. P. Loria, *Biochemistry* **41,** 6072 (2002).

[77] N. A. Farrow, O. W. Zhang, J. D. Formankay, and L. E. Kay, *J. Biomol. NMR* **4,** 727 (1994).

[78] A. G. Palmer, C. D. Kroenke, and J. P. Loria, *Methods Enzymol.* **339,** 204 (2001).

[79] H. Desvaux and P. Berthault, *Prog. NMR Spectrosc.* **35,** 295 (1999).

[80] V. A. Daragan and K. H. Mayo, *Prog. NMR Spectrosc.* **31,** 63 (1997).

[81] J. G. Kempf and J. P. Loria, "Meth. Mol. Biol.: Protein NMR Techniques," Vol. 278. Humana Press, Totowa, NJ, 2004.

[82] J. M. Schurr, B. S. Fujimoto, R. Diaz, and B. H. Robinson, *J. Magn. Reson.* **140,** 404 (1999).

Conformational Exchange (Flexibility) Gives Rise to NMR Chemical
  Shift Exchange

Even though biological macromolecules access a continuum of conformational states, many biological processes are known to behave in a two-
state manner where the macromolecule predominantly populates two
distinct conformational states. Examples of such two-state processes include enzyme allostery (active and inactive states), ligand binding (bound
and free states), and the folding of many small proteins (folded and
unfolded states). This chapter will consider only cases of interconversion
between two conformational states, however, formalisms for treatment of
cases that adopt multiple states have been developed and are a generalization of the limited two-state case discussed herein.[57,83]

Interconversion of an atom or molecule between two distinct conformational states involves molecular motion that almost always results in that
atom or molecule experiencing two magnetically distinct environments.
The consequence is a dramatic impact on the intensity, line width, and
chemical shift of the NMR signal. To aid the discussion, consider the case
in which the macromolecule does not undergo exchange on the microsecond to millisecond time scale. For this case, a single NMR signal is observed with an intensity that is proportional to the population. The line
width of the NMR signal is proportional to R_2, the transverse relaxation
rate constant, which is a measure of the lifetime of the signal in the so-
called transverse plane (orthogonal to the static magnetic field, B_0). The
chemical shift of the NMR signal is governed by the distinct chemical
environment experienced by a nucleus.

When a nucleus experiences the presence of conformational flexibility
on the microsecond to millisecond time scale, all three of these observables
are affected (Fig. 1). Specifically, each state gives rise to a distinct NMR
chemical shift. The difference in these chemical shifts is $\Delta\omega$ (in rad s^{-1}),
and reports on the difference in chemical environment of each state. Thus,
the interconversion between two states, or conformational exchange, gives
rise to what is formally known as NMR chemical shift exchange, which is
namely the exchange between two distinct NMR chemical shifts. Note that
chemical shift exchange could also refer to a chemical kinetic process, such
as hydrogen–deuterium exchange,[84,85] but is not considered herein.

Conformational exchange almost always gives rise to NMR chemical
shift exchange and affects the NMR signal. The rate constant that governs

[83] A. Allerhand and E. Thiele, *J. Chem. Phys.* **45,** 902 (1966).
[84] Y. W. Bai, J. J. Englander, L. Mayne, J. S. Milne, and S. W. Englander, *Methods Enzymol.*
  259, 344 (1995).
[85] G. Hernandez and D. M. LeMaster, *Magn. Reson. Chem.* **41,** 699 (2003).

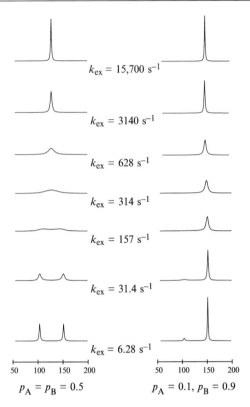

FIG. 1. Conformational exchange between two states gives rise to NMR chemical shift exchange. The left panel depicts chemical shift exchange for equal populations at various values of the exchange rate constant, k_{ex}. The right panel depicts skewed populations $p_A = 0.1, p_B = 0.9$. The spectra were simulated $R^0_{2A} = R^0_{2B} = 5$ s^{-1} and $\Delta\omega = 314$ rad s^{-1} using Eq. (2) from Huang and Oas.[69]

the chemical shift exchange process is k_{ex}, and is related to the conformational exchange rate constants depending on the type of motion involved as detailed in Eqs. (1–3). NMR chemical shift exchange causes phase decoherence of the NMR signal in the transverse plane resulting in an increase in the transverse relaxation rate constant, R_2, relative to the R_2 in the absence of exchange (called R^0_2). The effect is reflected in the lineshape but also impacts the observed chemical shift (Fig. 1).

The observed NMR signal, therefore, depends upon the transverse relaxation rate in the absence of chemical shift exchange, R^0_2, the chemical shift difference between the two states, $\Delta\omega$, the rate constant for chemical shift exchange, k_{ex}, and the populations of the two states, p_A and p_B.

Consider the case in which two conformational states are equally populated. In the event that chemical shift exchange is much slower than the difference in chemical shift (slow chemical shift exchange limit, $k_{ex} << \Delta\omega$), one observes two peaks in the NMR spectrum, at the chemical shift of each state. In the event that chemical shift exchange is much faster than the difference in chemical shift (fast chemical shift exchange limit, $k_{ex} >> \Delta\omega$), one observes a single peak in the NMR spectrum at the average chemical shift of the two states. The so-called intermediate chemical shift exchange limit is defined as when $k_{ex} \approx \Delta\omega$, and results in a single broad peak at the average of their chemical shifts. The time scale for NMR chemical shift exchange, therefore, is defined in part by the difference between the chemical shifts. Thus, a biological process that undergoes conformational exchange described by a single rate constant for the entire macromolecule actually could give rise to individual residues that undergo either slow, intermediate, or fast chemical shift exchange, because the NMR chemical shift time scale is governed not only by the k_{ex} of the overall process but also the chemical shift differences of those residues ($\Delta\omega$).

For many biological processes of interest, the populations of each state will be highly skewed. Consider these examples where $k_{ex} = k'_1 + k_{-1}$ and k'_1 is the pseudo-first-order reaction rate constant:
Protein folding or allostery case:

$$A_{\text{inactive}} \underset{k_{-1}}{\overset{k_1}{\rightleftharpoons}} A_{\text{active}} \qquad k'_1 = k_1 \tag{1}$$

Ligand or macromolecular binding case:

$$\text{A} + \text{L} \underset{k_{-1}}{\overset{k_1}{\rightleftharpoons}} \text{A} \cdot \text{L} \qquad k'_1 = k_1[\text{L}] \tag{2}$$

Homooligomerization case:

$$n\text{A} \underset{k_{-1}}{\overset{k_1}{\rightleftharpoons}} \text{A}_n \qquad k'_1 = nk_1[\text{A}]^{n-1} \tag{3}$$

In each case, typically one species will strongly predominate at a protein concentration suitable for NMR spectroscopy. Therefore, the effect of skewed populations is also illustrated in Fig. 1. By contrast to the case in which each state is equally populated, when conformational exchange gives rise to populations that are skewed the presence and time scale of exchange are difficult to detect. Thus, the observation of a single resonance in the NMR spectrum does not always indicate the presence of fast chemical shift exchange[68]; a single resonance is also consistent with two other situations: (1) slow exchange with highly skewed populations, and (2) no exchange whatsoever. We discuss below a strategy to differentiate these situations.

In summary, conformational exchange gives rise to changes in three observables in the NMR signal. These three observables depend upon six parameters as dictated by Eq. (4),[57,59,62] the complexity of which prevents one from readily identifying the relative contributions of each parameter to the NMR signal. One goal of this chapter is to demonstrate how these parameters affect the observed NMR signal.

$$R_2(1/\tau_{cp}) = \frac{1}{2}\left(R_a + R_b + k_{ex} - \frac{1}{\tau_{cp}}\cosh^{-1}[D_+\cosh(\eta_+) - D_-\cos(\eta_-)]\right)$$

(4)

where $D_\pm$ and $\eta_\pm$ are both functions of p_A and p_B, R_{2A}^0, R_{2B}^0, $\Delta\omega$, and k_{ex} as depicted elsewhere.[59,62]

Is Exchange Present?

How then might the investigator determine the presence and time scale of exchange for such cases with skewed populations? The answer lies in determining R_{ex}, the contribution to the observed transverse relaxation rate constant (R_2) that arises from chemical shift exchange:

$$R_2(\tau_{cp}) = R_2^0 + R_{ex}$$

(5)

where R_2^0 is the transverse relaxation rate constant in the absence of exchange and $R_2(\tau_{cp})$ is written in this fashion to underscore the τ_{cp} dependence on R_2 as discussed in the following. Methods for indirectly determining R_{ex} are well known and include lineshape analysis[86] and model free analysis of spin relaxation data.[87] Here the discussion focuses on a class of recently developed NMR experiments that allow a more direct measurement of the R_{ex} contribution to $R_2(\tau_{cp})$.

Perhaps the most experimentally accessible methods for $R_2(\tau_{cp})$ determination are spin echo-based methods—both Carr–Purcell spin echoes with the Meiboom–Gill enhancement (CPMG) and Hahn echoes.[65,77,81,88–90] If a nucleus undergoes a chemical shift exchange event during the spin echo

[86] J. Sandstrom, "Dynamic NMR Spectroscopy." Academic Press, London, 1982.

[87] R. Cole and J. P. Loria, *J. Biomol. NMR* **26,** 203 (2003).

[88] F. A. A. Mulder, N. R. Skrynnikov, B. Hon, F. W. Dahlquist, and L. E. Kay, *J. Am. Chem. Soc.* **123,** 967 (2001).

[89] F. A. Mulder, B. Hon, D. R. Muhandiram, F. W. Dahlquist, and L. E. Kay, *Biochemistry* **39,** 12614 (2000).

[90] N. R. Skrynnikov, F. A. A. Mulder, B. Hon, F. W. Dahlquist, and L. E. Kay, *J. Am. Chem. Soc.* **123,** 4556 (2001).

delay (the τ_{cp} delay), the chemical shift exchange leads to phase decoherence that is reflected in an increase in $R_2(\tau_{cp})$ (Fig. 2). The more chemical shift exchange events during the τ_{cp} delay, the greater the contribution to $R_2(\tau_{cp})$. Therefore, if chemical shift exchange is present, then longer values of τ_{cp} will result in an increase of $R_2(\tau_{cp})$ when compared to $R_2(\tau_{cp})$ measured with shorter values of τ_{cp}. If chemical shift exchange is absent, then the $R_2(\tau_{cp})$ measured has no dependence on the length of τ_{cp} delay. In summary, one can determine if R_{ex} is contributing to $R_2(\tau_{cp})$ by comparing the values of $R_2(\tau_{cp})$ derived from the shortest and longest possible values of τ_{cp} $[\Delta R_2(\tau_{cp}) > 0]$.[78]

The experiment is run in a manner similar to measuring R_2 in that a single τ_{cp} delay is selected and the total duration of the CPMG period is then varied. The resulting data are processed and the volume of the NMR signal is plotted as a function of total duration of the CPMG period. The data are fit using a nonlinear least-squares routine to a single exponential and the transverse relaxation rate constant, R_2, is obtained for that

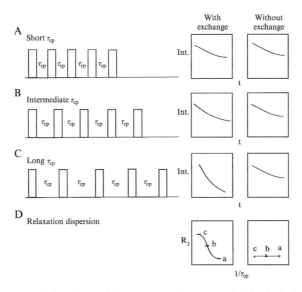

FIG. 2. Exchange during the τ_{cp} delay increases the observed $R_2(\tau_{cp})$. An increase in the τ_{cp} delay (A–C) increases the number of exchange events that occur between each CPMG $180°$ radiofrequency pulses, thereby leading to more phase decoherence and increased R_2 relaxation (faster decay of intensity versus time). For each value of τ_{cp}, a single value of $R_2(\tau_{cp})$ is determined and then plotted as a function of τ_{cp}^{-1}. The presence of chemical shift exchange results in a dispersion in the plot that is not observed in the absence of exchange. This relaxation dispersion curve (D) can be fit using Eq. (4) (or variations thereof) to determine R_2^0, p_A, k_{ex}, and $\Delta\omega$.

value of τ_{cp} from the exponential fit of the data.[91] A second value of τ_{cp} is then selected and data are once again collected as a function of the duration of the CPMG period. From these data, R_2 is obtained for the second value of τ_{cp}. Any difference in the values of $R_2(\tau_{cp})$ at the different τ_{cp} values indicates the presence of exchange $[\Delta R_2(\tau_{cp}) > 0]$.[78,92]

What is the Time Scale of Exchange?

Once the presence of exchange is confirmed the question of the time scale of chemical shift exchange arises. As noted earlier, residues involved in biological processes such as loop opening that are governed by identical kinetic rate constants can give rise to chemical shift exchange on different time scales because the NMR chemical shift time scale is defined not only by the rate constant, k_{ex}, but also by the chemical shift difference, $\Delta\omega$. Analysis of the relaxation dispersion data by curve-fitting to Eq. (4) can be used, in principle, to determine whether chemical shift exchange is slow or fast on the NMR chemical shift time scale.[93] However, as illustrated by Palmer and co-workers for the protein bovine pancreatic trypsin inhibitor, relaxation dispersion data at one static magnetic field strength do not always accurately define the NMR chemical shift time scale for exchange.[93] In their analysis, R_{ex} data collected at one static magnetic field strength demonstrated that residues Cys-38 and Ala-40 fit well to a limiting form of Eq. (4) that is valid only for fast exchange. However, these residues were actually determined to be undergoing slow chemical shift exchange when data collected at two static magnetic field strengths were analyzed. This observation illustrates that relaxation dispersion data collected at a single static magnetic field strength might not distinguish between slow and fast time scales.[62,93]

To define the NMR chemical shift time scale, therefore, one must collect relaxation dispersion data at different static magnetic field strengths. To rigorously determine the NMR chemical shift time scale requires the determination of the scaling parameter α, which reflects the R_{ex} dependence upon the static magnetic field strength[93]:

$$\alpha = d \ln \Delta R_{ex}/d \ln \Delta\omega \qquad (6)$$

where $0 \le \alpha < 1$ slow exchange
        $\alpha = 1$ intermediate exchange
        $1 < \alpha \le 2$ fast exchange

[91] A. M. Mandel, M. Akke, and A. G. Palmer, *J. Mol. Biol.* **246**, 144 (1995).

[92] A. G. Palmer, M. Akke, and A. M. Mandel, *Abstr. Pap. Am. Chem. Soc.* **214**, 165-PHYS (1997).

[93] O. Millet, J. P. Loria, C. D. Kroenke, M. Pons, and A. G. Palmer, *J. Am. Chem. Soc.* **122**, 2867 (2000).

The evaluation of α requires a wide range of τ_{cp} values in order to fully define the relaxation dispersion curve at a minimum of two different static magnetic field strengths, a procedure that consumes considerable instrument time. Fortunately, a reasonable approximation to the scaling parameter α is to measure relaxation dispersion using the very longest value of τ_{cp} at two static magnetic field strengths, and the very shortest value of τ_{cp} at only the lower strength of the two fields.[78] For ^{15}N, typical long and short values for τ_{cp} for this purpose are 10 and 1 ms, respectively. From these measurements, one can calculate α':

$$\alpha' = d \ln \Delta R_2(\text{long}\tau_{cp}, \text{short}\tau_{cp})/d \ln \Delta\omega \qquad (7)$$

$$\approx \left\{ (B_0^{hi} + B_0^{lo})/(B_0^{hi} - B_0^{lo}) \bullet (\Delta R_2^{hi} - \Delta R_2^{lo})/(\Delta R_2^{hi} + \Delta R_2^{lo}) \right\} \qquad (8)$$

where
$$\Delta R_2^{hi} = [R_2^{hi}(\text{long}\tau_{cp}) - R_2^{lo}(\text{short}\tau_{cp})]$$

$$\Delta R_2^{lo} = [R_2^{lo}(\text{long}\tau_{cp}) - R_2^{lo}(\text{short}\tau_{cp})]$$

As discussed by Millet et al.[93] a value of parameter $\alpha' < 1$ always indicates slow exchange, whereas a value of $\alpha' > 1$ could arise from either slow or fast exchange.

Data Analysis

The goal of this section is to discuss important considerations for data analysis. While significant issues surround data collection, two excellent reviews have recently covered this important topic.[78,81] The application of the CPMG-based techniques to study chemical exchange of a process can be done in a systematic way, and is delineated into a series of steps as detailed in the following.

Step 1. Simulation of exchange—Computer simulation of the relaxation dispersion curve for the process under study.

Step 2. Confirmation of exchange—Initial set of NMR experiments at two τ_{cp} values to determine the presence of exchange in the microsecond to millisecond time scale.

Step 3. Time scale of exchange—Determine whether the chemical shift exchange is slow or fast by determining the scaling parameter α as discussed previously.

Step 4. k_{ex} determination—Determine the exchange rate constant by fitting to the appropriate expression for slow [Eq. (4)] or fast [Eq. (9)] exchange.

Step 1: Simulation of Exchange

The first step for an investigator interested in quantifying conformational exchange is to study the feasibility of the CPMG-based approach for their system. In general, an equilibrium process of exchange between two biological states in the microsecond to millisecond time scale could be studied using the CPMG-based methods. However, before using the CPMG-based methods to determine the chemical shift exchange-based parameters, it is helpful to simulate the predicted relaxation dispersion-based equations for the phenomenological rate constants [Eq. (4)]. Because Eq. (4) depends on k_{ex}, p_A, $\Delta\omega$ and R_2^0, a priori knowledge of an approximate value for these parameters is necessary to simulate the relaxation dispersion curve. In many cases, such knowledge is known or reasonable estimates can be inferred. For example, k_{ex} can be obtained from global kinetic studies of the exchange process, p_A from equilibrium studies, $\Delta\omega$ from NMR correlation experiments of the two biological states, and R_2^0 can be estimated based on the molecular weight of the system. The simulations will aid the investigator in determining the feasibility of a CPMG-based approach. In addition, the simulations provide a guideline for choosing optimal τ_{cp} values necessary to define the relaxation dispersion curve.

Chymotrypsin Simulation Example. The following example of catalysis of *N*-acetyl tyrosine ethyl ester by chymotrypsin is illustrative of the simulation process. One needs reasonable estimates of k_{ex}, $\Delta\omega$, p_A, and R_2^0. For k_{ex}, we assume that k_{cat} is the rate-limiting step and the step of interest. A value of $k_{cat} = 190$ s^{-1} has been reported. The value of R_2^0 used in the simulation does not have to be very accurate as it does not define the overall shape of the curve but does define the offset from the *x*-axis (Fig. 3). Thus, R_2^0 can be assumed to be about 5 s^{-1} for proteins under 10 kDa and about 10 s^{-1} for larger ones around 25–30 kDa. If a slightly more rigorous estimate of R_2^0 is desired, it can be obtained using the Stokes–Einstein relation and the relationship that governs how the transverse relaxation rate varies with the τ_c for an isotropically rotating spherical macromolecule (the spectral density functions).[94] For example, the molecular weight of chymotrypsin is 25.1 kDa giving a rotational correlation time of 8.9 ns assuming a spherical shape and a 3 Å shell of water at 298 K. Using this value and assuming typical values for ^{15}N-labeled proteins including common backbone flexibility ($S^2 \approx 0.85$), then the R_2^0 is calculated to be 8.3 s^{-1} at 600 MHz. To determine the expected population

[94] J. Cavanagh, W. J. Fairbrother, A. G. Palmer, and N. J. Skelton, "Protein NMR Spectroscopy: Principles and Practice." Academic Press, San Diego, 1996.

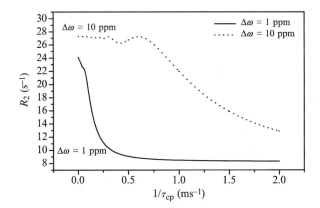

FIG. 3. The catalysis of *N*-acetyl tyrosine ethyl ester by chymotrypsin will give rise to relaxation dispersion. Simulation of Eq. (4) using expected parameters for chymotrypsin as discussed in the text. Note the sensitivity of the simulations to $\Delta\omega$.

(of the ES complex) at NMR concentrations, we estimate a value of 0.1 given the K_M of 6.6×10^{-4} M reported for chymotrypsin.[95] For the simulation a value of $\Delta\omega$ is also needed and could be obtained readily from an HSQC correlation spectrum of chymotrypsin in the free form as well as bound to substrate. If the value of $\Delta\omega$ is not known a priori, the simulation could still be performed with the typical values for $\Delta\omega$ (~1 to 10 ppm for ^{15}N chemical exchange). Using these values in Eq. (4), the relaxation dispersion curve is then simulated (Fig. 3). Both the curves show appreciable relaxation dispersion with a value of ΔR_2 of ~5 s^{-1} (for longest and shortest τ_{cp} values). The folding of the small protein chymotrypsin inhibitor C12 is also simulated using data obtained from the literature[96] (Fig. 4). In this simulation, the sensitivity of the CPMG-based methods to small populations is demonstrated by varying p_A between extreme values (99%, 90%, and 50% folded) at $\Delta\omega$ values of 1 and 10 ppm. Thus, the folding rates for this protein could be determined by CPMG-based methods if the unfolded state were populated.

Step 2: Confirmation of Exchange

If exchange occurs in the appropriate time scale for a CPMG-based approach, then the next step would be to experimentally confirm the presence of chemical shift exchange by measuring $R_2(\tau_{cp})$ at two extrema for τ_{cp},

[95] D. Voet and J. G. Voet, "Biochemistry." John Wiley & Sons, New York, 1995.
[96] T. R. Killick, S. M. Freund, and A. R. Fersht, *FEBS Lett.* **423,** 110 (1998).

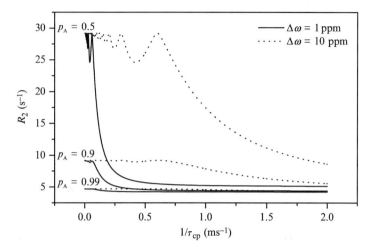

FIG. 4. The folding of chymotrypsin inhibitor 2 will give rise to relaxation dispersion. Simulation of Eq. (4) using data obtained from the literature.[96] Note the sensitivity of the simulations to the population of the minor species.

short and long.[38,78,93] $\Delta R_2(\tau_{cp})$ is calculated as the difference in $R_2(\tau_{cp})$ at the two τ_{cp} values and plotted against residue number. The presence of significant nonzero values indicates the presence of chemical shift exchange on the time scale that could be studied using CPMG-based experiments.

Step 3: Time Scale of Exchange

How does one then proceed to determine the NMR chemical shift time scale? In practice, the investigator should measure α' as described previously [Eqs. (7)–(8)]. If $\alpha' < 1$ then the slow time scale regime is unambiguously determined, and analysis of the system requires full relaxation dispersion data only at a single static magnetic field strength. If $\alpha' > 1$, the chemical shift time scale for exchange is still ambiguous. The investigator must collect full relaxation dispersion data at a minimum of two static magnetic field strengths to differentiate between slow and fast exchange.[78]

Step 4: k_{ex} Determination

For quantitative determination of chemical exchange, the transverse relaxation rate constant, $R_2(\tau_{cp})$, must be determined at enough values of τ_{cp} to fully define the relaxation dispersion curve. The appropriate number of τ_{cp} values to measure depends upon the chemical shift exchange time scale, which dictates the shape of the dispersion curve (e.g., Fig. 4). A minimum of six to eight values (not including duplicate measurements for error analysis) is necessary, but more values of τ_{cp} will increase the accuracy of the k_{ex} determination.

For each value of τ_{cp}, one needs to determine the transverse relaxation rate constant, $R_2(\tau_{cp})$. This determination is straightforward: The duration of the CPMG period (at a single value of τ_{cp}) is varied with time and the resulting crosspeak volumes fit to an exponential decay with a rate constant defined as R_2 using a nonlinear least-squares routine.[97,98] The uncertainties in the rate constants should be determined as discussed by Mandel et al.[99] The number of time points necessary to fully define the exponential decay of transverse signal is an important consideration that has been discussed elsewhere.[78,81,98,99] An important aspect of this consideration is to collect more data at the shorter and longer time points as these data are more useful in restraining the fit. In fact, Jones recently has proposed that only two time points, short and long extrema, are required, although the application of this approach to CPMG-based data has not been reported.[100]

Once values of R_2 have been determined at different τ_{cp} delays, these values are plotted as a function of $1/\tau_{cp}$ and the resulting curve is fit to Eq. (4) [or a limiting case of Eq. (4)]. For slow chemical shift exchange ($\alpha' < 1$), values of $R_2(\tau_{cp})$ from a single static magnetic field strength should be fit to Eq. (4). If $\alpha' > 1$, then the chemical shift time scale is ambiguous and multiple field data are necessary to define the time scale for exchange. The multiple field data can then be globally fit to Eq. (4). If the resulting analysis indicates that the fast exchange ($k_{ex} > \Delta\omega$), then significant improvements in the uncertainty of the data analysis can be achieved by fitting to the fast-limit form derived by Luz and Meiboom that has one less fitting variable than Eq. (4)[60]:

$$R_2(1/\tau_{cp}) = R_2(1/\tau_{cp} \to \infty) + R_{ex}\left[1 - \frac{2\tanh\left(\dfrac{k_{ex}\tau_{cp}}{2}\right)}{(k_{ex}\tau_{cp})}\right] \qquad (9)$$

For the curve fitting, two shareware programs have been developed that are well documented. The software program CPMGFit developed by Palmer interfaces UNIX-based shell scripts with the XMGR program for

[97] W. H. Press, B. P. Flannery, S. A. Teukolsky, and W. T. Vetterling, "Numerical Recipes in Fortran." Cambridge University Press, Cambridge, 1992.
[98] J. H. Viles, B. M. Duggan, E. Zaborowski, S. Schwarzinger, J. J. A. Huntley, G. J. A. Kroon, H. J. Dyson, and P. E. Wright, J. Biomol. NMR 21, 1 (2001).
[99] A. M. Mandel, M. Akke, and A. G. Palmer, J. Mol. Biol. 246, 144 (1995).
[100] J. A. Jones, J. Magn. Reson. 126, 283 (1997).

curve fitting and plotting.[101] CPMGFit features the ability to fit R_{ex} data to Eq. (4) and various limiting forms. CPMGFit also provides several options for estimating the uncertainties in the fitted parameters including covariance matrix analysis, Monte Carlo simulations, and jackknife simulations.[102] The second shareware option for data analysis is the software NMRView,[103] which contains an R_{ex} data analysis module that has been contributed by Mueller and others.[104]

Sensitivity Analysis for Eq. (4)

How does R_{ex} depend upon the populations of states (p_A), the transverse rate constant in the absence of exchange (R_2^0), the exchange rate constant (k_{ex}), and the chemical shift difference ($\Delta\omega$)? The sensitivity of each of these four variables of Eq. (4) is shown by simulations of experimental relaxation dispersion data collected for the $^{13}C_\alpha$ of Leu-6 of the *de novo* designed protein α_2D.[105,106] This protein is a dimeric four-helix bundle that has all the hallmarks of a native-like protein.[107–110] α_2D undergoes conformational exchange between an unfolded monomer and a folded dimer species.[108] Thus the chemical shift exchange contributions to the NMR resonances of α_2D arise from its folding; the rate constants of which were determined from global analysis of relaxation dispersion data collected at $B_0 = 11.7\ T$ and 14.1 T.[105] In the simulations presented here, three of the four parameters described in Eq. (4) are held constant at the value reported from the global analysis of α_2D. The fourth parameter is also held constant at different variables to illustrate how sensitive the data are to this parameter.

Not surprisingly, the simulations indicate that the experimental data are very sensitive to the transverse relaxation rate constant in the absence of exchange, R_2^0 (Fig. 5). As expected, the dependence of $R_2(\tau_{cp})$ on the τ_{cp} delay is independent of R_2^0 as illustrated by the invariant shape of the

[101] A. G. Palmer III, http://cpmcnet.columbia.edu/dept/gsas/biochem/labs/palmer/software/cpmgfit.html
[102] F. Mosteller and J. W. Tukey, "Data Analysis and Regression. A Second Course in Statistics." Pearson Education, Essex, 1977.
[103] B. A. Johnson and R. A. Blevins, *J. Biomol. NMR* **4**, 603 (1994).
[104] http://dir.niehs.nih.gov/dirnmr/scripts/
[105] R. B. Hill, C. Bracken, W. F. DeGrado, and A. G. Palmer, *J. Am. Chem. Soc.* **122**, 11610 (2000).
[106] R. B. Hill, D. P. Raleigh, A. Lombardi, and W. F. DeGrado, *Acc. Chem. Res.* **33**, 745 (2000).
[107] D. P. Raleigh, S. F. Betz, and W. F. Degrado, *J. Am. Chem. Soc.* **117**, 7558 (1995).
[108] R. B. Hill and W. F. Degrado, *J. Am. Chem. Soc.* **120**, 1138 (1998).
[109] R. B. Hill and W. F. Degrado, *Struct. Fold. Des.* **8**, 471 (2000).
[110] R. B. Hill, J. K. Hong, and W. F. Degrado, *J. Am. Chem. Soc.* **122**, 746 (2000).

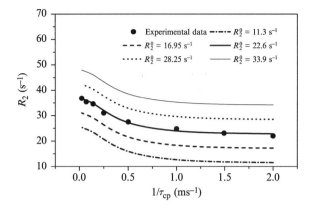

FIG. 5. Sensitivity of R_2^0 to fitting relaxation dispersion data at 11.7 T for $^{13}C_\alpha$ of Leu-6 of α_2D. The solid line depicts the best fit to the data from a global analysis of R_{ex} data collected at 11.7 and 14.1 T.[105] The other lines depict simulations using an R_2^0 value held constant at 50, 75, 125, and 150% of the best fit value. Values of $p_M = 0.038$, $\Delta\omega = 838.38$ rad s^{-1}, and $k_{ex} = 510$ s^{-1} were also held constant for each simulation.

curves at different values of R_2^0. Based on this analysis, the uncertainty in determining the transverse relaxation rate constant in the absence of exchange is less than 5%. Jackknife simulations performed on the α_2D data indicated the uncertainty at 2.3%.[105]

A small change in the population of the minor species dramatically affects the relaxation dispersion (Fig. 6). In the case for α_2D, detectable contributions to R_{ex} are possible even with chemical shift exchange to an unfolded state that is less than 1% populated. At the smaller values of τ_{cp}^{-1} (longer τ_{cp} delays), the population differences have a more pronounced effect as expected because more exchange events are occurring during the longer delay period. This pattern is also observed for the other two variables (k_{ex} and $\Delta\omega$) as expected.

The change in chemical shifts between the exchanging states ($\Delta\omega$) also strongly affects the fits to the experimental data as expected (Fig. 7). Note that the change in chemical shifts is inextricably linked to the populations of the exchanging sites as illustrated in Fig. 1 and governed by Eq. (4). Thus, R_{ex} measurements at only one static magnetic field strength will not allow one to determine both the populations and the change in chemical shifts. However, these parameters can be determined if R_{ex} measurements are made at more than one magnetic field strength, which constrains the value for $\Delta\omega$. In contrast, simulations are less sensitive to k_{ex} (Fig. 8). For measurements at only one static magnetic field strength, k_{ex} can be determined within ∼25–30% accuracy. However, the accuracy

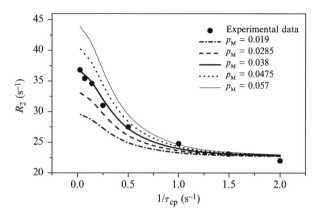

FIG. 6. Sensitivity of p_M to fitting relaxation dispersion data at 11.7 T for $^{13}C_\alpha$ of Leu-6 of α_2D. The solid line depicts the best fit to the data from a global analysis of R_{ex} data collected at 11.7 and 14.1 T.[105] The other lines depict simulations using a p_M value held constant at 50, 75, 125, and 150% of the best fit value. Values of $R_2^0 = 22.6$ s^{-1}, $\Delta\omega = 838.38$ rad s^{-1}, and $k_{ex} = 510$ s^{-1} were also held constant for each simulation.

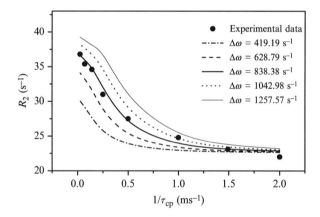

FIG. 7. Sensitivity of $\Delta\omega$ to fitting relaxation dispersion data at 11.7 T for $^{13}C_\alpha$ of Leu-6 of α_2D. The solid line depicts the best fit to the data from a global analysis of R_{ex} data collected at 11.7 and 14.1 T.[105] The other lines depict simulations using a $\Delta\omega$ value held constant at 50, 75, 125, and 150% of the best fit value. Values of $R_2^0 = 22.6$ s^{-1}, $p_A = 0.038$, and $k_{ex} = 510$ s^{-1} were also held constant for each simulation.

can be significantly improved by increasing the number of data points collected or global analysis of data collected at multiple static magnetic field strengths.

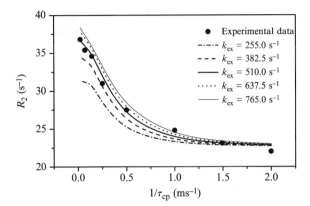

FIG. 8. Sensitivity of k_{ex} to fitting relaxation dispersion data at 11.7 T for $^{13}C_{\alpha}$ of Leu-6 of α_2D. The solid line depicts the best fit to the data from a global analysis of R_{ex} data collected at 11.7 and 14.1 T.[105] The other lines depict simulations using a k_{ex} value held constant at 50, 75, 125, and 150% of the best fit value. Values of $R_2^0 = 22.6$ s^{-1}, $p_A = 0.038$, and $\Delta\omega = 838.38$ rad s^{-1} were also held constant for each simulation.

The previous examples illustrate the exquisite dependence of the relaxation dispersion on R_2^0, p_M, k_{ex}, and $\Delta\omega$. If a priori knowledge of R_2^0, p_M, and $\Delta\omega$ is possible, then these parameters can be fixed in the fitting procedure and tightly restrain values for the rate constant k_{ex}. Is such knowledge possible? Indeed it is for the cases considered here. For example, in ligand binding the R_2^0 in the absence of exchange is readily determined from NMR spin relaxation experiments of the unliganded macromolecule. In the case of protein allostery this might not be possible, however, experiments with enhanced sensitivity to chemical exchange allow the explicit determination of R_2^0.[67] The populations of the species (p_A) is typically available from other biophysical measurements such as fluorescence binding assays in the case of ligand binding and chemical or thermal denaturation in the case of protein folding. The change in chemical shift between the exchanging states ($\Delta\omega$) can sometimes be determined independently in many systems.

How important is a priori knowledge of Eq. (4) variables in obtaining reasonable fits to experimental R_{ex} data? While knowledge of the transverse relaxation rate constant in the absence of exchange, R_2^0, can improve the fits; the differences in the fits when R_2^0 is fixed at its known value or when it is allowed to float are similar if a fully relaxed dispersion curve is possible to obtain experimentally. However, for some systems this is not the case and measurable dispersion is still evident even at the shortest values of τ_{cp} (long τ_{cp}^{-1}). For these cases, a priori knowledge of R_2^0 will significantly reduce the uncertainty in the fits to experimental data.

Limitations of CPMG-Based Methods

The time scale of exchange events that are measurable by CPMG-based methods ranges from 100 s^{-1} to 3000 s^{-1} depending on whether $\Delta\omega$ is sufficiently large. For these cases, fitting Eq. (4) to experimental data to obtain all four parameters can result in large uncertainties for k_{ex}, $\Delta\omega$, and p_A because of the multiple combinations of these parameters that still gives reasonable fits to the data. Furthermore, at a single magnetic field strength it is impossible to separate the populations of the exchanging species from their chemical shift differences. However, several strategies allow the investigator to overcome these obstacles. First, independent determination of R_2^0, $\Delta\omega$, and p_A is often possible. In the cases in which such data are not available, then global analysis of data collected at multiple magnetic field strengths can be used to improve the accuracy of the fits. Finally, the exchange equilibrium can be manipulated in some manner (e.g., temperature, mutagenesis, inhibitors, or substrate analogues) to affect the change of some of the fitting parameters without affecting others. Subsequent global analysis of all data can yield acceptable fits to the data.

Conclusion

In the past few years, the tools for measuring microsecond to millisecond time scale motions by NMR spectroscopy have been developed. These tools combine the great strengths of NMR: the abilities to probe molecular motions in a site-specific manner and to study dynamic processes. In the future, the wide application of these methods to important biological problems will have a huge impact. The traditional notion of biological structure will be replaced with a more realistic view of fluctuating conformational substates, some of which are biologically active. The judicious application of the methods discussed herein, coupled with biochemical, biophysical, and molecular dynamics simulations, will be essential for identifying the active substates from the inactive ones. These studies will ultimately reveal how conformational flexibility is related to biological activity.

Author Index

Numbers in parentheses are footnote reference numbers and indicate that an author's work is referred to although the name is not cited in the text.

Subject Index

A

Analysis of variance, *see* Repeated-measures analysis of variance

ANOVA, *see* Analysis of variance; Mixed-model regression analysis

Approximate entropy
definition, 172
sample entropy comparison, 182–184
statistical bias of analysis, 173–174

B

Blood glucose monitoring, *see* Glycosylated hemoglobin

C

Coupling strength, *see* Hormone concentration coupling, polynomial transfer functions

D

Deconvolution analysis, hormone pulse detection
advantages, 53–54
burst area evaluation, 50
CLUSTER algorithm comparison with PULSE4, 51–53
luteinizing hormone concentration simulation, 47–50
Monte Carlo simulations, 46–47
peak-detection performance, 49–50
PULSE4 algorithm, 40, 43–46
sampling interval effects, 50–51
secretion profile, 40–43

Diabetes, *see also* Glycosylated hemoglobin
complications, 95
epidemiology, 94–95
glycemic control optimization, 95–96, 104–105
monitoring, 96–97
types, 94

Differential equation set matrix exponentiation, *see* Matrix exponentiation

E

Endocrine network modeling
concentration dynamics simulation for single hormones, 56–60
hormone concentration coupling, *see* Hormone concentration coupling, polynomial transfer functions
multiple feedback loop networks, 75–78, 81
principles, 54–56, 78–79
pulse detection, *see* Hormone pulse, deconvolution analysis in detection
single system feedback loop oscillations
formal 2-node/1-feedback network, 60–61, 79
node identification and oscillation control, 71–72, 81
oscillations generated by a periodic solution, 64–65
oscillations generated by a perturbation, 69–71
reference systems, 61–64, 80
simulations
antibody infusion, 65–66, 80
exogenous infusion, 67, 69
sensitivity modification, 66–67
synthesis separation from secretion, 73–75

Entropy, *see* Approximate entropy; Sample entropy analysis

F

Fourier approximations, *see* Wavelet analysis

G

GlobalWorks program, matrix exponentiation, 25–28

Glucose monitoring, *see* Glycosylated hemoglobin

275

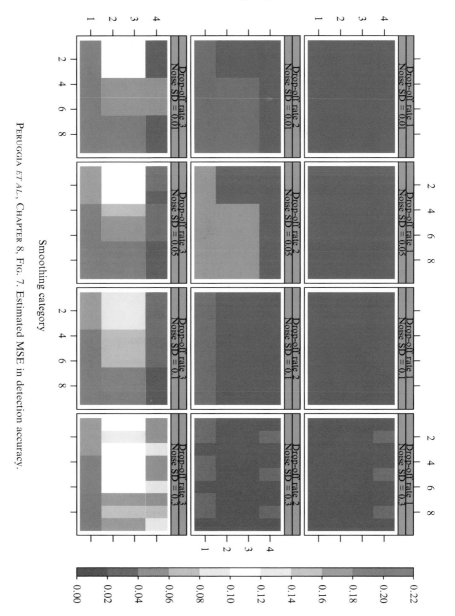

PERUGGIA ET AL., CHAPTER 8, FIG. 7. Estimated MSE in detection accuracy.

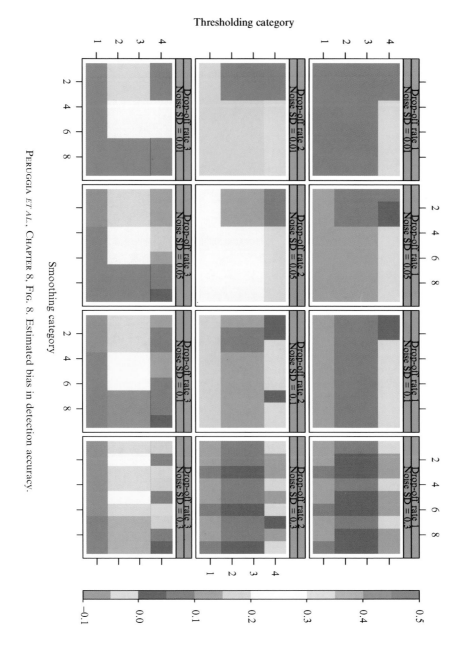

PERUGGIA *ET AL.*, CHAPTER 8, FIG. 8. Estimated bias in detection accuracy.